T0296202

CAMBRIDGE MONOGRAPHS ON PHYSICS

GENERAL EDITORS

N. FEATHER, F.R.S.
Professor of Natural Philosophy in the University of Edinburgh

D. SHOENBERG, PH.D.
Fellow of Gonville and Caius College, Cambridge

SOME RECENT RESEARCHES IN SOLAR PHYSICS

SOME RECENT RESEARCHES IN SOLAR PHYSICS

BY

F. HOYLE, M.A.
Fellow of St. John's College, Cambridge

CAMBRIDGE
AT THE UNIVERSITY PRESS
1949

CAMBRIDGE
UNIVERSITY PRESS

University Printing House, Cambridge CB2 8BS, United Kingdom

Published in the United States of America by Cambridge University Press, New York

Cambridge University Press is part of the University of Cambridge.

It furthers the University's mission by disseminating knowledge in the pursuit of education, learning and research at the highest international levels of excellence.

www.cambridge.org
Information on this title: www.cambridge.org/9781107418929

© Cambridge University Press 1949

This publication is in copyright. Subject to statutory exception and to the provisions of relevant collective licensing agreements, no reproduction of any part may take place without the written permission of Cambridge University Press.

First published 1949
First paperback edition 2014

A catalogue record for this publication is available from the British Library

ISBN 978-1-107-41892-9 Paperback

Additional resources for this publication at www.cambridge.org/9781107418929

Cambridge University Press has no responsibility for the persistence or accuracy of URLs for external or third-party internet websites referred to in this publication, and does not guarantee that any content on such websites is, or will remain, accurate or appropriate.

GENERAL PREFACE

The Cambridge Physical Tracts, out of which this series of Monographs has developed, were planned and originally published in a period when book production was a fairly rapid process. Unfortunately, that is no longer so, and to meet the new situation a change of title and a slight change of emphasis have been decided on. The major aim of the series will still be the presentation of the results of recent research, but individual volumes will be somewhat more substantial, and more comprehensive in scope, than were the volumes of the older series. This will be true, in many cases, of new editions of the Tracts, as these are re-published in the expanded series, and it will be true in most cases of the Monographs which have been written since the War or are still to be written.

The aim will be that the series as a whole shall remain representative of the entire field of pure physics, but it will occasion no surprise if, during the next few years, the subject of nuclear physics claims a large share of attention. Only in this way can justice be done to the enormous advances in this field of research over the War years.

N. F.
D. S.

May, 1948.

CONTENTS

CHAPTER VII

The emission of radio waves from the sun

AUTHOR'S PREFACE

This tract was written during the summer months of 1947. Re-reading the manuscript a year later I find, especially in the last two chapters, that on several issues I have been guilty of too confident assertion, while, on the other hand, a number of suggestions have been strengthened by subsequent investigation. In spite of the temptation to rewrite the passages concerned, I have decided to adhere to the original form. New considerations are added as brief supplementary notes at the end of Appendix II.

In compiling the sections dealing with observational material I have frequently had recourse, in addition to papers cited in the text, to articles by G. Abetti and S. A. Mitchell in the *Handbuch der Astrophysik* and also to *Physik d. Sternatmosphären* by A. Unsöld. The work of Chapters III and IV, which contain the bulk of the original material included in the tract, was carried out in collaboration with R. A. Lyttleton and H. Bondi (*M.N.* **107**, 184, 1947). Although, in the remainder of the tract, the amount of theoretical material taken directly from astronomical literature is not large, my views on many important questions have been much influenced by other authors. In particular, my outlook on electromagnetic effects at and below the photosphere has arisen largely from the work of S. Chapman and T. G. Cowling, while the discussion of electromagnetic effects in the solar atmosphere has been mainly stimulated by the investigations of R. G. Giovanelli. I am indebted to Dr. Giovanelli for allowing me to look over preliminary drafts of several recent papers. My thanks are also due to D. R. Bates, who has been my guide in all matters relating to the Earth's atmosphere.

F. H.

ST. JOHN'S COLLEGE,
CAMBRIDGE.
27 *June*, 1948

CHAPTER I

SUNSPOTS AND THE SOLAR CYCLE

1. The Photosphere

When viewed directly in projection against the sky the limb of the sun appears sharp. This is due to a very rapid decrease of emission in the continuum with increasing distance from the solar centre. At a distance $\sim 6{\cdot}9 \times 10^{10}$ cm., where the density of material $\sim 10^{-8}$ g./cm.^{3}, and the temperature $\sim 5740°$ K., the emission in all spectral lines together is small compared with the emission in the continuum, but, at a level a few hundred kilometres further from the centre, the continuous emission falls to a value comparable with the emission in $H\alpha$. For descriptive purposes it is convenient to regard this change as occurring discontinuously, and we refer to the level in question as the *photosphere.* Material above the photosphere will be described as belonging to the *solar atmosphere.* In the present chapter we are concerned with phenomena occurring at, and below, the photosphere.

2. Individual Sunspots and Sunspot Groups

(i) *The formation and lifetimes of sunspots*

Quite apart from sunspots, the surface of the sun is not smooth. There is a general speckled appearance† or *granulation*, due to bright nuclei, with diameters less than 1000 km., which probably arise from convection occurring in layers below the photosphere. Although the nuclei are usually separated by distances of the order of their diameters, there are often spaces where no nuclei are visible. If such spaces increase in size they become black and are then called *pores.* Sunspots are formed by the coalescence of a number of pores, and have effective surface temperatures $\sim 4500°$ K.

The lifetimes of sun-spots range from a few hours up to about two months. The dependence of spot area on time for a long-lived spot is shown in Fig. 1.

† H. H. Plaskett, *M.N.*, **96**, 402, 1936.

(ii) *The appearance and size of sunspots*

Fully developed spots, when near the centre of the solar disk, are observed to be of roughly circular form with a dark inner region or *umbra*, surrounded by a less dark annulus or *penumbra*. The change from umbra to penumbra is remarkably sudden. The umbra occupies about a fifth of the total area.

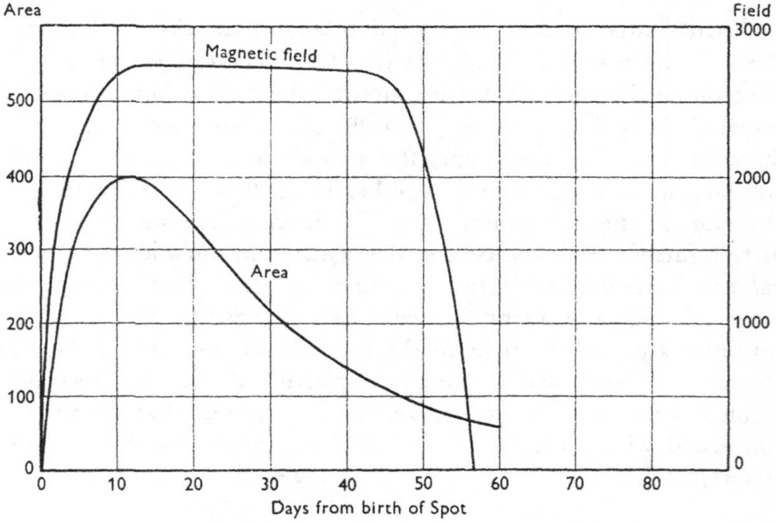

Fig. 1. Area in millionths of the sun's visible hemisphere, and central field in gauss.

Observation at the solar limb suggests that the surface of a spot has the form of a shallow funnel with centre depressed below the outer rim of the penumbra by about one tenth of the radius of the spot.

A typical spot has radius ~10,000 km., but exceptional spots with radii as large as 50,000 km. are occasionally observed. Very large spots are usually distorted in shape.

(iii) *The grouping of sunspots*

Although both single spots and groups containing more than two are observed, there is a strong tendency for spots to occur in pairs. The two spots of a pair lie approximately on the same circle of solar latitude, and the preceding spot, in the sense of the solar rotation, is often called the leader of the pair. The separation of a spot pair is of the same order as the diameters of the spots themselves, but is variable owing to a tendency for the spots to drift apart, particularly during the early stages in their formation.

(iv) *Magnetic properties*

The discovery by Hale of the magnetic fields of sunspots has led to important discoveries throughout solar physics. For sim-

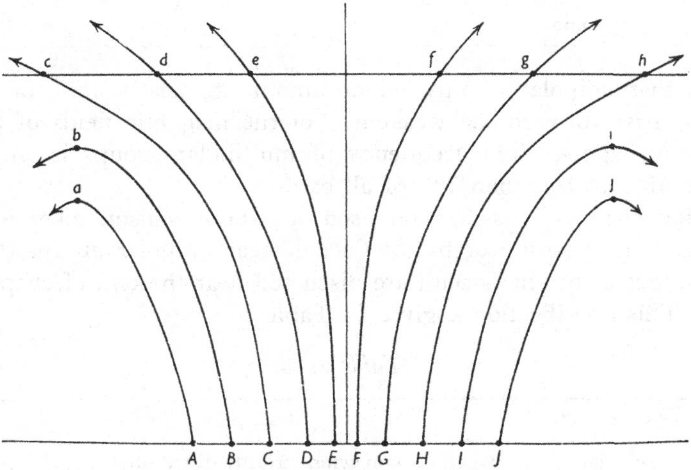

Fig. 2. Lines of magnetic force in a vertical plane through the axis of a sunspot. AJ is the level of the base of the Eddington zone. *ch* is the level of the photosphere.

plicity, we take the field to be axially symmetric about the solar radius through the centre of the spot. A form of the field, consistent with observation, is shown in Fig. 2. The variation of the surface field throughout the lifetime of a long-lived spot is given in Fig. 1.

The maximum field attained in the life of a spot depends on the maximum radius of the spot. This is shown in Table 1. Although fields as high as 4000 gauss sometimes occur in large spots, the magnetic field tends to saturate at about 3000 gauss.

Both magnetic polarities are observed. In particular, a spot pair is often called a bipolar group because the two spots show opposite polarities. About 90% of all sunspot groups are of bipolar character. There is a strong tendency for the field associated with the leader to be greater than the field of the follower. This suggests that unipolar groups, which amount to nearly 10% of all cases, arise through the weakening of the magnetic fields of the following spots. The frequency of multipolar groups is small, amounting to less than 1% of all cases.

Table 1.

Radius of penumbra (km.)	Central surface magnetic intensity (gauss)
3000	500
5000	1000
8000	1500
11,000	2000
15,000	2500
> 20,000	~ 3000

The magnetic classification used at Mount Wilson takes into account the position of bright flocculi near unipolar groups (the properties of bright flocculi are discussed near the end of chapter II). This classification is given in Table 2.

Table 2.

Magnetic Character	Description
(α) Unipolar	Bright flocculi symmetrically distributed
(αp) Unipolar	Centre of spot group precedes bright flocculi
(αf) Unipolar	Centre of spot group follows bright flocculi
(β) Bipolar	Spot areas for each polarity approximately equal
(βp) Bipolar	The preceding component is the principal member
(βf) Bipolar	The following component is the principal member
($\beta\gamma$) Bipolar	Either the preceding or the following component is accompanied by smaller components of opposite polarity
(γ) Multipolar	Contain spots of both polarities, but distributed too irregularly to be described as bipolar

(v) *The Evershed effect*

The material at the surface of a spot is observed to be flowing out of the spot with an average velocity ~2 km./sec. This outflow probably follows the lines of magnetic force which become tangential at the edge of the spot. There is no definite observational evidence in favour of vortical motion in sunspots.

Table 3.—Observed Sunspot Relative Numbers
1749–1946

Year	R.N.	Year	R.N.	Year	R.N.	Year	R.N.	Year	R.N.
1749	80·9	1789	118·1	1829	67·0	1869	73·9	1909	43·9
1750	**83·4**	1790	89·9	1830	**71·0**	1870	**139·1**	1910	18·6
1751	47·7	1791	66·6	1831	47·8	1871	111·2	1911	5·7
1752	47·8	1792	60·0	1832	27·5	1872	101·7	1912	3·6
1753	30·7	1793	46·9	1833	8·5	1873	66·3	1913	1·4
1754	12·2	1794	41·0	1834	13·2	1874	44·7	1914	9·6
1755	9·6	1795	21·3	1835	56·9	1875	17·1	1915	47·4
1756	10·2	1796	16·0	1836	121·5	1876	11·3	1916	57·1
1757	32·4	1797	6·4	1837	**138·3**	1877	12·3	1917	**103·9**
1758	47·6	1798	4·1	1838	103·2	1878	3·4	1918	80·6
1759	54·0	1799	6·8	1839	85·8	1879	6·0	1919	63·6
1760	62·9	1800	14·5	1840	63·2	1880	32·3	1920	37·6
1761	**85·9**	1801	34·0	1841	36·8	1881	54·3	1921	26·1
1762	61·2	1802	45·0	1842	24·2	1882	59·7	1922	14·2
1763	45·1	1803	43·1	1843	10·7	1883	**63·7**	1923	5·8
1764	36·4	1804	**47·5**	1844	15·0	1884	63·5	1924	16·7
1765	20·9	1805	42·2	1845	40·1	1885	52·2	1925	44·3
1766	11·4	1806	21·1	1846	61·5	1886	25·4	1926	63·9
1767	37·8	1807	10·1	1847	98·5	1887	13·1	1927	69·0
1768	69·8	1808	8·1	1848	**124·3**	1888	6·8	1928	**77·8**
1769	**106·1**	1809	2·5	1849	95·9	1889	6·3	1929	65·0
1770	100·8	1810	0·0	1850	66·5	1890	7·1	1930	35·7
1771	81·6	1811	1·4	1851	64·5	1891	35·6	1931	21·2
1772	66·5	1812	5·0	1852	54·2	1892	73·0	1932	11·1
1773	34·8	1813	12·2	1853	39·0	1893	**84·9**	1933	5·7
1774	30·6	1814	13·9	1854	20·6	1894	78·0	1934	8·7
1775	7·0	1815	35·4	1855	6·7	1895	64·0	1935	36·1
1776	19·8	1816	**45·8**	1856	4·3	1896	41·8	1936	80·4
1777	92·5	1817	41·1	1857	22·8	1897	26·2	1937	**113·5**
1778	**154·4**	1818	30·4	1858	54·8	1898	26·7	1938	106·4
1779	125·9	1819	23·9	1859	93·8	1899	12·1	1939	89·8
1780	84·8	1820	15·7	1860	**95·7**	1900	9·5	1940	66·4
1781	68·1	1821	6·6	1861	77·2	1901	2·7	1941	50·5
1782	38·5	1822	4·0	1862	59·1	1902	5·0	1942	30·4
1783	22·8	1823	1·8	1863	44·0	1903	24·4	1943	15·3
1784	10·2	1824	8·5	1864	47·0	1904	42·0	1944	11·0
1785	24·1	1825	16·6	1865	30·5	1905	**63·5**	1945	33·2
1786	82·9	1826	36·3	1866	16·3	1906	53·8	1946	92·6
1787	**132·0**	1827	49·7	1867	7·3	1907	62·0	1947	~151·0
1788	130·9	1828	62·5	1868	37·3	1908	48·5		

3. The Solar Cycle

(i) *Relative sunspot numbers*

The Wolfer relative sunspot numbers from 1749 are given in Table 3. These numbers, which are based on an empirical system, are not exactly proportional to the surface area covered by spots. Nevertheless, they give a satisfactory index of the variation from year to year in solar activity. More precise data, obtained at Greenwich, shows that for example at the 1917 maximum, 1537 millionths of the solar surface was covered by spots. Although the proportionality between the relative numbers and actual areas varies somewhat with the phase of the solar cycle (particularly near sunspot minimum) this case may be taken as giving an approximate factor of proportionality for the whole cycle.

(ii) *The sunspot period*

The mean sunspot period is about 11·1 years. The average interval from minimum to maximum is only 4·5 years, as compared with an average of 6·6 years from maximum to minimum.

(iii) *An asymmetry of distribution*

Spots were almost entirely absent from the northern hemisphere during the prolonged period of minimum activity at the end of the eighteenth and at the beginning of the nineteenth centuries. In a number of other cycles there has also been an appreciable difference, in one direction or the other, between the spot frequencies in the two hemispheres of the sun.

(iv) *The latitude drift*

Sunspots are not only periodic in number, but also show a periodic variation in latitude. The spots of a new cycle break out approximately in solar latitudes 30° N. and 30° S. These belts drift towards the equator as the sunspot number increases until they reach about 16° N. and 16° S. at sunspot maximum. The drift towards the equator continues, but now with decreasing sunspot number. The spots finally die out at about 8° N. and 8° S. Two or three years before the final disappearance of these spots new disturbances break out again in latitudes 30° N. and 30° S., and the cycle is repeated. Whereas at sunspot maximum there are only two belts of spots, at minimum there are four belts; two near the equator and two in higher latitudes.

(v) *The reversal of magnetic polarities*

The bipolar sunspot groups are systematically orientated in each hemisphere. In one hemisphere the magnetic polarity of the leader spots is always *N* and the followers *S*, and in the other hemisphere the polarities are reversed; the leader spots being *S* and the followers *N*. The orientation is preserved throughout each sunspot cycle, but, with outbreak at sunspot minimum of new spots in latitudes 30° N. and 30° S., the polarities are all reversed. Thus if, for example, the polarity of the leader spots in the northern hemisphere is *S* in a given cycle, then in the following cycle it will be *N*. An analysis at Mount Wilson of a large number of spot groups has shown that exceptions to this rule occur in only about 3% of all cases. Finally, it may be noted that nothing is known of the magnetic characteristics of sunspots before 1908. Thus the observations on which the above remarks are based are necessarily limited to the last four solar cycles.

4. Some General Properties

(i) *The rotation of the sun*

The period of rotation of the sun at different latitudes has been obtained from observations of long-lived spots. The values given by Carrington are shown in Table 4. The equatorial rotational

Table 4.

Latitude (°N and °S)	Rotation period (days)
0	25·0
10	25·2
20	25·7
30	26·5

velocity ~2 km./sec. Estimates can also be obtained from other long-lived markings on the solar surface, such as bright flocculi. Although the values so obtained differ slightly from those given in Table 4, they confirm the theoretically important conclusion that the sun rotates most rapidly at the equator.

(ii) *The general magnetic field of the sun*

Between 1912 and 1914, Hale, Seares, van Maanen, and Ellerman obtained evidence for a general solar magnetic field by measuring the polarization in the wings of a number of spectral lines. The field was ~50 gauss and possessed an axis of symmetry inclined to the solar axis of rotation by ~5°. Recently Thiessen has also found evidence for a general field ~50 gauss.

The measurements involved in this work are of a very delicate nature, and both Hale and Thiessen have expressed doubts concerning the validity of their conclusions. As, however, there is strong indirect evidence (see chapter v) in support of a general field, the existence of this field will be assumed throughout the following investigations.

5. Theoretical Discussion

(i) *Bjerknes' description of the solar cycle*

It is supposed that each belt of sunspots arises from a vortex tube extending round the sun along a parallel of latitude. The vortex tube is regarded as producing a spot whenever it intersects the photosphere. Such intersections evidently occur in pairs, and the vortical motion of the photospheric material in two adjacent intersections must be in opposite senses.

It is further supposed that the magnetic fields of sunspots arise from the vortical motion, and the polarity is taken as determined by the sense of this motion. The different magnetic orientations of bipolar spots in the northern and southern hemispheres is taken to be due to the corresponding vortex tubes in the two hemispheres being in opposite senses.

To complete the set of hypotheses a zonal vortex in each hemisphere is assumed first to approach the photosphere in latitude 30° and then to drift towards the equator. When the vortex reaches latitude 8°, after a drift lasting 11·1 years, a second tube with opposite vorticity appears at latitude 30°. The second tube then drifts towards the equator while the first tube falls to deeper layers and is convected back to higher latitudes. That is, in each hemisphere two tubes of opposite vorticity alternate with each other in a cycle that consists in drifting (at no great depth below the photosphere) from latitude 30° to latitude 8°, then in falling to deeper layers, and finally in being convected back to latitude 30°, where the photosphere is again approached.

These suggestions of Bjerknes show how the main features of solar activity can be described by four assumptions: (*a*) the existence and drift of the vortex tubes, (*b*) the connexion of the magnetic fields with vortical motion, (*c*) a periodicity of 11·1 years for the drift from 30° to 8°, and (*d*) the alternation in each hemisphere of two tubes of opposite vorticity. Later work has shown that the difficulties involved in satisfying these requirements are very great. In particular, it seems unlikely that vorticity can be as important as was originally supposed by Hale and Bjerknes.

For the moment we confine attention to the properties of individual spots. In chapter v the general questions concerning the solar cycle will be raised again.

(ii) *The darkening of sunspots*

It was suggested by Russell that a sunspot acts as a pumping agency that lifts material from the base of the spot to the photosphere. The darkening is attributed to the absorption of energy that must occur as the rising material acquires gravitational potential energy. A quantitative discussion † of the process leads to an estimate ~10,000 km. for the depth of the spot.

The motion of the material is taken to be initially vertical, but later becomes approximately horizontal as the photosphere is approached. This horizontal motion, which appears as the Evershed effect, leads to material being deposited in regions surrounding the spot. Thus, although the potential energy must finally be converted back to thermal energy and radiation, the reconversion takes place outside the spot.

An important criticism has recently been advanced by Cowling, who points out that the sharp change from umbra to penumbra is not explained by this theory. To overcome this objection it is necessary to consider the effect of the magnetic field of the spot on the convective process. Chapman ‡ has drawn attention to the fact that the magnetic field of a sunspot is of such a character (curl $\boldsymbol{H} \neq 0$) that momentum must be communicated by the field to the material within the spot. Thus if $\boldsymbol{j}$ is the current density and $\boldsymbol{H}$ the magnetic field there is a body force $\boldsymbol{j} \times \boldsymbol{H}/c$ per unit volume, where c is the velocity of light, acting on the material. This force can be written entirely in terms of the electric field $\boldsymbol{E}$ and the magnetic field, by using the equation (we put both the dielectric constant and the permeability of the material equal to unity)

$$4\pi \boldsymbol{j} = c \operatorname{curl} \boldsymbol{H} - \frac{\partial \boldsymbol{E}}{\partial t}. \qquad (5{\cdot}1)$$

† A. Unsöld, *Physik d. Sternatmosphären*, J. Springer, 1938, p. 387.
‡ S. Chapman, *M.N.*, **103**, 117, 1943.

Then Euler's equations of motion for the material can be expressed as

$$(5\cdot 2) \qquad \frac{\partial \boldsymbol{u}}{\partial t} + \tfrac{1}{2}\,\text{grad}\,\boldsymbol{u}^2 - \boldsymbol{u} \times \text{curl}\,\boldsymbol{u} = -\frac{1}{4\pi\rho}\,.\,\boldsymbol{H} \times \left(\text{curl}\,\boldsymbol{H} - \frac{1}{c}\frac{\partial \boldsymbol{E}}{\partial t}\right) - \frac{1}{\rho}\,\text{grad}\,P,$$

where ρ, $\boldsymbol{u}$ are the mass-density and velocity respectively of the material, P is the hydrostatic pressure, and t represents time. It is shown below, in part (iv) of this section, that the term in $\boldsymbol{H} \times \partial \boldsymbol{E}/\partial t$ is only rarely important in solar physics. But the term in $\boldsymbol{H} \times \text{curl}\,\boldsymbol{H}$ may be important, when, as in a sunspot, at least one of the components of curl $\boldsymbol{H}$ is of the same order as the space derivatives of $\boldsymbol{H}$.

In a sunspot, as Chapman pointed out, the $\boldsymbol{H} \times \text{curl}\,\boldsymbol{H}$ term tends to produce a spreading of the magnetic lines of force. The quasi-stability of the spot means that this tendency must be overcome by an excess hydrostatic pressure outside the spot. This is made possible by the cooling in the spot. It is a tautology, however, to claim † that this necessity, for a lower temperature within the spot, explains why the cooling occurs. The relative importance in (5·2) of the $\boldsymbol{H} \times \text{curl}\,\boldsymbol{H}$ and grad P terms can be estimated, so far as order of magnitude is concerned, by comparing P with the magnetic energy $\boldsymbol{H}^2/8\pi$ per unit volume. The later work of this subsection suggests that P exceeds 10^7 dynes/cm.^{2} near the base of a spot. With such a pressure, it is possible to hold a spot together, provided $|\boldsymbol{H}|$ is not much greater than 10^4 gauss. Near the photosphere, however, P falls to a value of order 10^4 dynes/cm.^{2}, and the magnetic term must there predominate. Thus, near the photosphere, we expect an expansion of both material and of the magnetic lines of force. This effect is capable of explaining the Evershed effect, and also of why the lines of force fan out as the photosphere is approached.

The fanning out of the magnetic lines of force near the photosphere can be incorporated in a convective theory by using Eddington's discussion of the hydrogen convection zone.‡ According to Eddington this zone extends from immediately below the photosphere down to a depth (probably ~10,000 km.)

† H. Alfén, *M.N.*, **105**, 4, 1945. ‡ A. S. Eddington, *M.N.*, **102**, 154, 1942.

where the pressure $\sim 1 \cdot 7 \times 10^7$ dynes/cm.^{2}, and the temperature $\sim 27{,}000°$ K. It follows, since gravity g is $2 \cdot 74 \times 10^4$ cm. sec.$^{-2}$ at the photosphere, that the amount of material in the convection zone ~ 620 gr./cm.^{2} The material is almost wholly hydrogen; the total contribution of all the metals being less than about 1% by mass.

The feature of Eddington's theory of particular importance in the sunspot problem is that within the convection zone the energy flux is mainly by convection and not by radiation. That is, radiation is converted near the base of the zone into energy transported by convection, and the latter energy flux is converted back to radiation as upward moving material approaches the photosphere. It is important to notice that although there may be an excess transfer of material in the ascending columns over the transfer in the descending columns (as Russell suggested), this is not a feature of Eddington's theory in which the convective flux arises from an *excess of ionization* of hydrogen in the ascending material.

Under normal conditions the convection currents are mainly directed along solar radii, but as shown in part (iv) of the present section the directions of flow in a sunspot must lie essentially along the magnetic lines of force. Then owing to the fanning out of the lines of force the upward moving currents reach the photosphere over an area considerably greater than the base from which they started. This is illustrated in Fig. 2, where currents moving upwards from a circular base of diameter CH reach the photosphere over a circular area of diameter ch. Such a deflection of the convection currents leads to the area covered by a spot losing, to regions outside, an appreciable part of the energy flux carried by convection. Hence, the emission at the surface of a spot must be markedly smaller than the normal rate of radiation at the photosphere.

The efficiency of convection in Eddington's theory depends on the process working between a lower level where hydrogen is largely ionized and an upper level where the hydrogen is almost wholly unionized. The latter condition requires the convection to continue up to a level near the photosphere. Thus the efficiency of convection must be seriously reduced for a line of force such as Aa, since a, which is at the highest working level on this line, is appreciably below the photosphere. Accordingly, the transport by convection decreases markedly outside a base of diameter *CH*. We take this base as defining the umbra of the spot.

So far, we have not invoked excess transfer of material in the ascending columns. The effect of the $\boldsymbol{H} \times \text{curl}\, \boldsymbol{H}$ term in (5·2), however, is to produce an outflow of material at the photosphere, and this outflow requires the amount of material moving upwards along the magnetic lines of force *Aa*, *Bb*, etc., to exceed the amount convected downwards along *aA*, *bB*, etc. This excess can explain the darkening outside the umbra. Thus we may regard the darkening of a sunspot as arising from the superposition of two effects:

(*a*) The process, suggested by Russell, of excess transfer of material in the upward convection currents produces a general darkening of the whole spot.

(*b*) The process, suggested by Eddington, of excess transfer of ionization energy in the upward convection currents, when taken in conjunction with the fanning out of the magnetic lines of force near the photosphere, produces an additional darkening over the umbra of the spot.

The above discussion depends on Eddington's theory of the convection zone. If this should prove erroneous it will probably be necessary to resort to radiative processes, in addition to convection. In this connexion a recent paper by Odgers † is of interest. He shows that a radiative diversion of energy occurs if the product of the absorption coefficient and density of material inside the spot is higher than in surrounding regions. By suitably adjusting the opacity a sharp change from umbra to penumbra can be obtained.

(iii) *The change of magnetic field in static material*

The $\partial \boldsymbol{E}/\partial t$ term in (5·1) can be neglected. Then the magnetic field $\boldsymbol{H}$ is given in terms of the current density $\boldsymbol{j}$ by the relation

$$\boldsymbol{H} = \frac{1}{c}\,\text{curl} \int \frac{\boldsymbol{j}\, d\tau}{r}, \tag{5·3}$$

when, as in the sunspot problem, there is no contribution to $\boldsymbol{H}$ from magnetic dipoles. In the present subsection the change of $\boldsymbol{H}$ in a sunspot is considered on the following assumptions:

† G. J. Odgers, *M.N.*, **106**, 101, 1946.

(1) The material in the spot is static.

(2) There is no initial distribution of electric charge.

(3) Both the magnetic field and the distribution of material in the spot are axially symmetric.

In the main body of a sunspot the magnetic field does not affect the relation between $\boldsymbol{j}$ and $\boldsymbol{E}$. This can be seen by considering two special cases. The general problem where $\boldsymbol{H}$ is neither perpendicular nor parallel to $\boldsymbol{E}$ can be considered as a combination of these cases.

(*a*) ***H*** *perpendicular to* ***E***.

The current $\boldsymbol{j}$ is given by †

(5·4) $$\boldsymbol{j} = \sigma^I \boldsymbol{E} + \sigma^{II} \boldsymbol{E} \times \boldsymbol{H}/|\boldsymbol{H}|,$$

(5·5) $$\sigma^I + i\sigma^{II} = c^2(kT^{-3/2} - i\,8{\cdot}6 \times 10^3 |\boldsymbol{H}|\,T/P_e)^{-1},$$

where P_e is the electron pressure in dynes/cm.², $i = \sqrt{-1}$, and k is a factor that depends in a complicated way on T, H, P_e. The variations in k are not large, however, and for most purposes it is sufficient to put $k = 6{\cdot}8 \times 10^{13}$. The 'transverse' conductivity σ^{II} is smaller than σ^I, and σ^I is not much reduced by the magnetic field, if

$$|\boldsymbol{H}| < 7{\cdot}9 \times 10^9\, P_e/T^{5/2} \text{ gauss.}$$

This condition gives $H < 6{\cdot}6 \times 10^5$ gauss in the case $P_e = 10^7$ dynes/cm.², $T = 27{,}000°$ K. (corresponding to Eddington's values for the base of the hydrogen convection zone). Thus in the main body of a spot we may neglect σ^{II} and, in accordance with (5·5), we take $\sigma^I/c^2 \sim 10^{-8}$, in agreement with an earlier estimate by Chapman.‡

(*b*) ***H*** *parallel to* ***E***.

The current in this case is given by

(5·4′) $$\boldsymbol{j} = \sigma \boldsymbol{E},$$

(5·5′) $$\sigma = c^2 T^{3/2}/k,$$

which agrees with (*a*) when, as in the main body of a sunspot, $|\boldsymbol{H}| < 7{\cdot}9 \times 10^9 P_e/T^{5/2}$ gauss.

We now follow closely a recent discussion by Cowling.§ By

† T. G. Cowling, *Proc. Roy. Soc.*, **183** A, 453, 1945.

‡ S. Chapman, *M.N.*, **89**, 56, 1928–9.

§ T. G. Cowling, *M.N.*, **106**, 218, 1946.

assumptions (1), (2), (3) the electric field is due solely to the change of $\boldsymbol{H}$ with time. For the current $\sigma' \boldsymbol{E}$ is directed round circles possessing the axis of symmetry as axis, and such a system does not lead to an accumulation of space charge. Thus $\boldsymbol{E}$ is given by

$$\boldsymbol{E} = -\frac{1}{c}\frac{\partial \boldsymbol{A}}{\partial t}, \tag{5·6}$$

where $\boldsymbol{A}$ is the magnetic vector-potential. Moreover we have

$$4\pi \boldsymbol{j} = 4\pi\sigma' \boldsymbol{E} = c \operatorname{curl} \boldsymbol{H} = -c\nabla^2 \boldsymbol{A} \tag{5·7}$$

in the interior of the spot. Equations (5·6) and (5·7) give

$$\nabla^2 \boldsymbol{A} = \frac{4\pi\sigma'}{c^2}\frac{\partial \boldsymbol{A}}{\partial t}. \tag{5·8}$$

Now $\nabla^2 \boldsymbol{A}$ cannot be less than a value $\sim \boldsymbol{A}/a^2$ for a spot of radius a. Thus by (5·8) the decay time of $\boldsymbol{A}$, and hence of $\boldsymbol{H}$, is at least $\sim\sigma' a^2/c^2$. Taking $a = 10{,}000$ km., $\sigma'/c^2 = 10^{-8}$, the time of decay of the magnetic field due to purely electromagnetic effects is at least ~ 300 years.

With assumptions (*a*), (*b*) and (*c*) the time of growth of a sunspot magnetic field due to a current generator must be of the same order. Suppose that currents begin to flow about the axis of the spot at time $t = 0$, and let these currents be such that they give rise to a vector-potential $\boldsymbol{A}_0$ after a sufficient time has elapsed for the field to attain a constant value. Then the total vector-potential $\boldsymbol{A}$ at any time is the sum of $\boldsymbol{A}_0$ and a part satisfying (5·8). That is

$$\nabla^2(\boldsymbol{A} - \boldsymbol{A}_0) = \frac{4\pi\sigma'}{c^2}\frac{\partial(\boldsymbol{A} - \boldsymbol{A}_0)}{\partial t}. \tag{5·9}$$

But $\nabla^2(\boldsymbol{A} - \boldsymbol{A}_0)$ cannot be appreciably greater than

$$(\boldsymbol{A} - \boldsymbol{A}_0)/a^2,$$

so that $\boldsymbol{A}$ cannot be appreciably greater than

$$\boldsymbol{A}_0\{1 - \exp(-\sigma' a^2 t/c^2)\}.$$

Taking $\sigma' a^2/c^2 = 300$ years, the field built up in 10 days cannot appreciably exceed one tenthousandth part of the steady field corresponding to the vector-potential $\boldsymbol{A}_0$.

The inference from the above discussion, when taken together with the observed rates of growth and decay of the magnetic fields of sunspots, is that the fields are not produced *in situ*. That is, we must now dismiss assumption (*a*), and consider the effect of the motion of material forming the spot.

(iv) *Magnetic fields in moving material*

When the material moves rectilinearly with uniform velocity the problem is elementary. In this case both the current j and the lines of magnetic force are carried along with the material. Thus any region of exceptionally high magnetic intensity moves with the common velocity of the system.

The general problem, in which velocity variations occur between different parts of the material, is complicated. But there is a comparatively simple case, of importance in the sunspot problem. If the material in a region of exceptionally high magnetic intensity moves *as a whole* relative to material in regions of much weaker intensity then, although there are intricate electromagnetic effects in the regions of weak intensity, the region of high intensity behaves essentially as in the elementary case described in the previous paragraph. The growth and decay of the magnetic fields of sunspots probably arises from the transport of magnetic energy due to this process. Thus the field can be built up through a region of high magnetic intensity rising to the photosphere from deeper layers.† Similarly, the field can decay by the material sinking back to deeper layers.

When applied to the motion of material within a sunspot, this discussion gives support to the assumption (see part (i) of the present section) that the material follows the magnetic lines of force. For appreciable velocities perpendicular to the lines of force would lead to a rapid transport of magnetic energy, which is in contradiction with the observed quasi-stability of the field.

The complexities that arise in the general case are illustrated by the following investigation of the influence of the magnetic field $\boldsymbol{H}$, produced by the material in a domain of high magnetic energy, on the material in a second separate domain. Since the domains are separate it follows that curl $\boldsymbol{H}$ is zero in the second domain. Hence, an electromagnetic body force acting on the material in the second domain can only arise from the $\boldsymbol{H} \times \partial \boldsymbol{E}/\partial t$ term in (5·2). Let the material in the second domain move with

† H. Alfén, *M.N.*, **105**, 3 and 382, 1945.

velocity $\boldsymbol{u}$ relative to the first domain. Then by applying a Lorentz transformation it can be shown that an observer moving with the second domain recognizes an electric field $\boldsymbol{u} \times \boldsymbol{H}/c$, due to the influence of the material in the first domain. Thus, again omitting σ^{II}, there is a current density $\sigma^{I}\boldsymbol{u} \times \boldsymbol{H}/c$. For the moment we neglect the electromagnetic field arising from this current distribution. Then the resulting force acting on the material of the second domain is $\sigma^{I}(\boldsymbol{u} \times \boldsymbol{H}) \times \boldsymbol{H}/c^2$ per unit volume. If $\boldsymbol{u}$ is parallel to $\boldsymbol{H}$ the force is zero, but if $\boldsymbol{u}$ is perpendicular to $\boldsymbol{H}$ there is a force of magnitude $\sigma^{I} \mid \boldsymbol{u} \mid H^2/c^2$ per unit volume opposing the motion. This force, *if it persisted*, would destroy the motion represented by $\boldsymbol{u}$ in a time $\sim\rho c^2/\sigma^{I}H^2$, where ρ is the density of the material. In all problems of interest in solar physics this gives a time which is only a small fraction of a second.

To decide whether the retarding force persists in actuality for sufficiently long to destroy the motion, it is seen by comparing the $\rho\partial\boldsymbol{u}/\partial t$ and $\boldsymbol{H} \times \partial\boldsymbol{E}/\partial t$ terms in (5·2) that the $\boldsymbol{H} \times \partial\boldsymbol{E}/\partial t$ term is unable to produce an appreciable change in $\boldsymbol{u}$ if

$$(5\cdot10) \qquad \rho \mid \boldsymbol{u} \mid \gg \mid \boldsymbol{H} \times \boldsymbol{E} \mid /4\pi c$$

throughout the motion. In the present problem $\boldsymbol{E} = \boldsymbol{u} \times \boldsymbol{H}/c$. Accordingly, the inequality (5·10) is satisfied if

$$(5\cdot11) \qquad \rho \gg H^2/4\pi c^2.$$

The latter condition holds in all cases considered in the present tract. It follows therefore that the retarding force does not persist long enough to have an appreciable effect on the motion in the cases of interest in the present work.

The argument of the preceding paragraph shows that the retarding force must be destroyed by the field produced by the flow of current arising from the motion of the material. Thus the *total* magnetic field, comprising $\boldsymbol{H}$ together with the field due to the current distribution, is such that the magnetic lines of force are carried along with the material (a slight velocity across the lines of force must still persist in order to produce the required modification of $\boldsymbol{H}$, but this is small compared with $\mid \boldsymbol{u} \mid$). Accordingly, an observer at rest relative to the field $\boldsymbol{H}$ moves with velocity close to $-\boldsymbol{u}$ relative to the total magnetic field.

An important feature in the modification of the magnetic field arises from the building of space charges. In the special case where curl ($\boldsymbol{u} \times \boldsymbol{H}$) is zero the charge distribution can be adjusted so as to produce a polarization field $-\boldsymbol{u} \times \boldsymbol{H}/c$. Then an observer moving with the material recognizes no electric field, whereas an observer at rest relative to the first domain recognizes the polarization field.

(v) *The change of the general solar magnetic field*

The growth and decay of the general magnetic field of the sun can be discussed † in a manner similar to the work of subsection (iii). The characteristic time is again found to be $\sim\sigma' R^2/c^2$, where the radius R ($6{\cdot}9 \times 10^{10}$ cm.) of the sun is used in place of the radius of the spot. The conductivity of material in the deep interior of the sun corresponds to $\sigma'/c^2 \sim 10^{-4}$, which gives a characteristic time $\sim 10^{10}$ years. A more detailed discussion of this question is given in Appendix II, subsection (i).

† T. G. Cowling, *M.N.*, **105**, 166, 1945.

CHAPTER II

THE CHROMOSPHERE AND CORONA: OBSERVATIONAL DATA

6. The Heights of the Chromosphere and Corona

The density under gravity g of a gas of mean molecular weight μ and uniform temperature T varies with height h according to the factor $e^{-h\mu g/\Re T}$. In the reversing layer $T = 4830°$ K., $g = 2{\cdot}74 \times 10^4$ cm. sec.$^{-2}$, and μ is close to unity on account of the great abundance of neutral hydrogen. (Strömgren has shown that † hydrogen atoms are about 10,000 times more numerous than the atoms of the commonest metals. At a temperature of 4830° K. the hydrogen is mainly neutral and $\mu \sim 1$.) Accordingly in the reversing layer the *scale height* $\Re T/g\mu \sim 146$ km.

The observations of Mitchell ‡ have been discussed by Wildt, § who finds, subject to certain assumptions (see section 20), that the concentration of the *excited* states of a given element can be taken, over a limited range of height, as depending on h through the factor $e^{-h/H}$. Table 5 gives numerical values for various elements.

These values of H are all greater than would occur under thermodynamic conditions in an atmosphere of scale height 146 km. But the discrepancy in the first part of Table 5 is small compared with that in the second part, where H exceeds the thermodynamic value corresponding to $T = 4830°$ K. by factors of order 10. The values of H increase sharply at a height ~2000 km. above the photosphere. Thus above 2000 km. conditions must be markedly non-thermodynamic so far as the density gradient of material is concerned, whereas below 1500 km. there is a rough approximation to thermodynamic values. We shall refer to the region from 1500 km. to about 12,000 km. as the *chromosphere*, and to the region from the photosphere to a height of 1500 km.

† B. Strömgren, *Festschrift für Elis Strömgren*, Copenhagen, 1940.

‡ S. A. Mitchell, *Ap.J.*, **105**, 1, 1947. § R. Wildt, *Ap.J.*, **105**, 36, 1947.

as the *reversing* layer. The distinction between these two regions is discussed in chapter IV.

The upper limits of the height ranges in Table 5 are based on the brightest lines of the elements concerned. For hydrogen, only $H\alpha$ extends to 12,000 km. Thus $H\beta$ extends to 9000 km., $H\gamma$ to 8000 km., etc., and $H37$, which is the last line observed in the Balmer series, extends to only 500 km. above the photosphere

Table 5.

Element	*Sc* II	*Ti* II	*V* II	*Cr* I	*Mu* II	*Fe* I	*Fe* II
Range of height above photosphere (km.)	800–1500	1000–1500	500–1200	500–1000	800–1500	500–1200	500–1200
Ionization potential (eV.)	12·8	13·6	14·1	6·74	15·70	7·83	16·5
H (km.)	270	230	240	210	310	270	240
Element	*Ni* I	*Y* II	*Zr* II	*Ca* I			
Range of height above photosphere (km.)	500–1000	500–1200	500–1200	500–1200			
Ionization potential (eV.)	7·61	12·3	13·97	6·09			
H (km.)	240	240	240	360			

Element	*Sc* II	*Ti* II	*Mg* I	*Ca* II	*Sr* II	*H*	*He*
Range of height above photosphere (km.)	1500–4500	2000–6000	2000–6000	10,000–14,000	1000–6000	500–12,000	2000–8000
Ionization potential (eV.)	12·8	13·6	7·61	11·82	10·98	13·53	24·46
H (km.)	~2000	~2000	870	1100	620	1100	~4000

(lines above $H37$ are smeared into a continuum by the Stark effect). The brightest lines of the Paschen series extend to a height of 6500 km. The only unionized metal atoms giving lines above ~2000 km. are *Mg* I and *Ca* I (which extend to ~4000 km.).

The large values of H occurring in the second part of Table 5 suggest that the kinetic temperature of material above 2000 km. may be appreciably greater than the temperature of the reversing layer. This has been confirmed by Redman,† who obtained a temperature ~30,000° K. by attributing the observed widths of the chromospheric hydrogen lines to thermal broadening. An essential feature of Redman's argument is that these widths cannot

† R. O. Redman, *M.N.*, **102**, 140, 1942.

$\sim$150,000 km. beyond $r = 1{\cdot}1$. As pointed out by Alfèn,† such a scale height requires the coronal material to possess a kinetic temperature $\sim 2{\cdot}10^6$ deg. K. (since the material is largely ionized hydrogen the mean molecular weight is close to 0·5). This remarkable result is confirmed by three independent arguments. First, it has been noted by Grotrian ‡ that in order to explain the absence of Fraunhofer lines in the spectrum of the corona at distances r less than about 1·25, it is necessary for the scattering electrons to possess velocities $\sim 7{\cdot}5 \times 10^8$ cm./sec. The kinetic temperature required to give a mean thermal velocity of this order is $\sim 10^6$ deg. K. Second, the widths measured by Lyot of the brighter coronal emission lines, when interpreted as the result of thermal broadening, also indicate a temperature $\sim 10^6$ deg. K. (The line widths are unlikely to be due solely to thermal broadening, since convection currents also make a contribution. Nevertheless, this method must give a correct order of magnitude.) Working on this basis, Waldmeier § has recently obtained temperatures up to $6{\cdot}55 \times 10^6$ deg. K. Third, the ionization potentials listed in Table 7 are so high as to demand temperatures of order 10^6 deg. K. (see section 7).

Fraunhofer lines appear in the spectrum of the corona between $r \sim 1{\cdot}5$ and $r \sim 3$. Following Grotrian, it is generally thought that the appearance of these lines arises from scattering by dust particles. But volatization due to solar radiation prevents dust particles occurring within $r = 3$, so that the particles cannot occur in the corona itself, but would have to lie in a column between the sun and the earth. At first sight it seems that such a process cannot explain the marked concentration of the Fraunhofer spectrum towards the sun, but this difficulty can be overcome if the scattering has a strong component in the forward direction. This question has been considered by Allen, ‖ who points out that diffraction produces such a preferential scattering.

† H. Alfén, *Arkiv. Mat. Astr. Fys.*, **27**A, No. 25, 1941.

‡ W. Grotrian, *Z.Ap.*, **3**, 220, 1931.

§ M. Waldmeier, *Astro. Mitt. d. Eidg. Sternwarte Zürich*, No. 149, 1947.

‖ C. W. Allen, *M.N.*, **106**, 137, 1946. In the writer's opinion certain numerical work in this paper is called into question by the assumption (foot of p. 146) that the intensity of the diffracted field is equal to the total energy falling on the particles. This much overestimates the diffracted field when the radius of the particle is large compared with the wavelength of the radiation.

Values of n_e for $r < 1{\cdot}1$ are also given, but are open to doubt as accurate measurements of intensity become difficult near the solar limb. We prefer to obtain the electron density in the chromosphere by the method given in section 8. The total intensity of scattered light from the whole of the corona is about one millionth of the luminosity of the sun ($3{\cdot}78 \times 10^{33}$ ergs/sec.).

Table 7.—Emission lines in the solar corona (after Edlén)

λ (Ang-ströms)	Relative intensity	Emitting atom	Transition prob. (sec.$^{-1}$)	Excitation potential (eV.)	Ionization potential (eV.)
3328	~ 1	*Ca* XII	488	3·72	589
3388·1	~ 16	*Fe* XIII	87	5·96	325
3454·1	~ 2·3				
3601·0	~ 2·1	*Ni* XVI	193	3·44	455
3642·9		*Ni* XIII	18	5·82	350
3800·8					
3986·9	~ 0·7	*Fe* XI	9·5	4·68	261
4086·3	~ 1·0	*Ca* XIII	319	3·03	655
4231·4	~ 2·6	*Ni* XII	237	2·93	318
4311					
4359		*A* XIV ?	108	2·84	682
4567	~ 1·1				
5116·03	~ 3·2	*Ni* XIII	157	2·42	350
5302·86	100	*Fe* XIV	60	2·34	355
5536		*A* X	106	2·24	421
5694·42	~ 1·2	*Ca* XV ?	95	2·18	814
6374·51	~ 13	*Fe* X	69	1·94	233
6701·83	~ 3·7	*Ni* XV	57	1·85	422
7059·62	~ 2·2	*Fe* XV		31·7	390
7891·94	~ 13	*Fe* XI	44	1·57	261
8024·21	~ 0·5	*Ni* XV	22	3·39	422
10746·80	~ 55	*Fe* XIII	14	1·15	325
10797·95	~ 35	*Fe* XIII	9·7	2·30	325

The corona shows an emission spectrum of an entirely different character from the chromospheric lines. Following a suggestion due to Grotrian, Edlén has obtained † the identification shown in Table 7 for these lines. The energy emitted in the lines is about 1% of the total scattered energy in the continuum. The intensities fall off with increasing r at approximately the same rate as $n_e(r)$, and at a total eclipse the brightest lines can be detected out to about $r = 2{\cdot}5$.

The values of $n_e(r)$ given in Table 6 suggest a scale height

† B. Edlén, *Z.Ap.*, **22**, 30, 1942, and *M.N.*, **105**, 323, 1945.

$\sim$150,000 km. beyond $r = 1{\cdot}1$. As pointed out by Alfèn,† such a scale height requires the coronal material to possess a kinetic temperature $\sim 2{\cdot}10^6$ deg. K. (since the material is largely ionized hydrogen the mean molecular weight is close to 0·5). This remarkable result is confirmed by three independent arguments. First, it has been noted by Grotrian ‡ that in order to explain the absence of Fraunhofer lines in the spectrum of the corona at distances r less than about 1·25, it is necessary for the scattering electrons to possess velocities $\sim 7{\cdot}5 \times 10^8$ cm./sec. The kinetic temperature required to give a mean thermal velocity of this order is $\sim 10^6$ deg. K. Second, the widths measured by Lyot of the brighter coronal emission lines, when interpreted as the result of thermal broadening, also indicate a temperature $\sim 10^6$ deg. K. (The line widths are unlikely to be due solely to thermal broadening, since convection currents also make a contribution. Nevertheless, this method must give a correct order of magnitude.) Working on this basis, Waldmeier § has recently obtained temperatures up to $6{\cdot}55 \times 10^6$ deg. K. Third, the ionization potentials listed in Table 7 are so high as to demand temperatures of order 10^6 deg. K. (see section 7).

Fraunhofer lines appear in the spectrum of the corona between $r \sim 1{\cdot}5$ and $r \sim 3$. Following Grotrian, it is generally thought that the appearance of these lines arises from scattering by dust particles. But volatization due to solar radiation prevents dust particles occurring within $r = 3$, so that the particles cannot occur in the corona itself, but would have to lie in a column between the sun and the earth. At first sight it seems that such a process cannot explain the marked concentration of the Fraunhofer spectrum towards the sun, but this difficulty can be overcome if the scattering has a strong component in the forward direction. This question has been considered by Allen, || who points out that diffraction produces such a preferential scattering.

† H. Alfén, *Arkiv. Mat. Astr. Fys.*, **27**A, No. 25, 1941.

‡ W. Grotrian, *Z.Ap.*, **3**, 220, 1931.

§ M. Waldmeier, *Astro. Mitt. d. Eidg. Sternwarte Zürich*, No. 149, 1947.

|| C. W. Allen, *M.N.*, **106**, 137, 1946. In the writer's opinion certain numerical work in this paper is called into question by the assumption (foot of p. 146) that the intensity of the diffracted field is equal to the total energy falling on the particles. This much overestimates the diffracted field when the radius of the particle is large compared with the wavelength of the radiation.

7. Ionization Equilibrium

The thermal motions of particles in the solar atmosphere are in local equilibrium (except in the special circumstances discussed in chapter v). This means that if n is the density of a particular type of particle, the number of such particles per unit volume with velocities lying between v and $(v + dv)$ is

$$(7{\cdot}1) \qquad 4\pi n\left(\frac{m}{2\pi kT}\right)^{3/2} e^{-mv^2/2kT} v^2 dv,$$

where T is the 'kinetic temperature' and m is the particle mass. For an assembly in thermodynamic equilibrium the energy density of radiation in the frequency range ν to $(\nu + d\nu)$ must be

$$(7{\cdot}2) \qquad \frac{8\pi h\nu^3}{c^3} \cdot \frac{d\nu}{e^{h\nu/kT} - 1}.$$

But in the chromosphere and corona this condition is not even approximately satisfied, except near the base of the chromosphere. Thus in the visible spectrum the value of T to be used in (7·2) is close to 5740° K. (the photospheric temperature) except at values of ν lying within the Fraunhofer lines, where a lower value of about 4000° K. must be employed. On the other hand the kinetic temperature exceeds 20,000° K. in the chromosphere and rises to 10^6 deg. K. in the corona. Accordingly, the concepts of thermodynamic equilibrium must be used with caution. This is illustrated by the following discussion of the ionization equilibrium of hydrogen, adopting first the assumption of thermodynamic equilibrium (that is, using the same value of T in (7·1) and (7·2), and second the assumption of zero energy density of radiation.

If n_p, n_n represent the densities of protons and neutral hydrogen atoms respectively, the ratio n_p/n_n under thermodynamic conditions is given by the Saha formula

$$(7{\cdot}3) \qquad \log\left(\frac{n_p P_e}{n_n}\right) = -x \, . \, \frac{5040}{T} + \frac{5}{2}\log T - 0{\cdot}48,$$

where P_e is the electron pressure in dynes/cm.², T is measured in °K., and x is here the ionization potential 13·53 eV. of the ground state of the hydrogen atom. The fraction $n_p/(n_p + n_n)$

of hydrogen that is ionized can be obtained from (7·3) when P_e is given.

Table 8.—Ionization of hydrogen under thermodynamic equilibrium

$T \backslash \log P_e$	6	4	2	0	−1	−2	−3
4200						$2·1 \times 10^{-6}$	$2·1 \times 10^{-5}$
4582					6×10^{-6}	6×10^{-5}	6×10^{-4}
5040				$1·7 \times 10^{-5}$	$1·7 \times 10^{-4}$	$1·7 \times 10^{-3}$	$1·7 \times 10^{-2}$
5600			5×10^{-6}	5×10^{-1}	5×10^{-3}	$4·8 \times 10^{-2}$	$3·34 \times 10^{-1}$
6300		$1·5 \times 10^{-6}$	$1·5 \times 10^{-4}$	$1·5 \times 10^{-2}$	$1·32 \times 10^{-1}$	$6·08 \times 10^{-1}$	$9·37 \times 10^{-1}$
7200		$4·8 \times 10^{-5}$	$4·8 \times 10^{-3}$	$3·24 \times 10^{-1}$	$8·27 \times 10^{-1}$	$9·80 \times 10^{-1}$	$9·98 \times 10^{-1}$
8400	$1·6 \times 10^{-5}$	$1·6 \times 10^{-3}$	$1·4 \times 10^{-1}$	$6·19 \times 10^{-1}$	$9·94 \times 10^{-1}$	$9·99 \times 10^{-1}$	
10080	$5·7 \times 10^{-4}$	$5·4 \times 10^{-2}$	$8·52 \times 10^{-1}$	$9·98 \times 10^{-1}$			

This tables gives $n_p/(n_p + n_n)$.

Next we turn to the case where the energy density of radiation is zero and examine the general problem of the ionization equilibrium between positive ions with $(Z+1)$-electrons stripped, and ions of ionization potential x with Z-electrons stripped (this general case will be applied to hydrogen by putting $Z = 0$, $x = 13·53$ eV.). The conversion of Z-ions to $(Z+1)$-ions is due to electron impact, the ionization taking place almost entirely from the ground state of the Z-ions. The ionization equilibrium is established as a balance between this process and the recombination of electrons with $(Z+1)$-ions. Thus in equilibrium the densities n_Z, n_{Z+1} of the ions satisfy the relation

(7·4)
$$\frac{n_Z}{n_{Z+1}} = \frac{\text{Recombination probability per unit time per } (Z+1)\text{-ion}}{\text{Ionization probability per unit time per } Z\text{-ion}}.$$

The cross-section for ionization of a Z-ion by an electron of energy $\eta(>x)$ is given approximately by

(7·5)
$$2·5 \times 10^{-38} l x^{-2} \sqrt{(\eta - x)/x},$$

where x is *now expressed in ergs*, and l is the number of electrons in the outermost shell. This cross-section taken together with the distribution (7·1) for the electrons leads to the expression

(7·6)
$$\frac{2·5 \times 10^{-38} l n_e}{x^{3/2}} \left(\frac{2}{m_e}\right)^{\frac{1}{2}} e^{-x/kT}$$

for the ionization probability per unit time per Z-ion, where m_e is the electronic mass.

If we treat the $(Z+1)$-ions as hydrogen-like the energy emitted in the frequency range ν to $(\nu+d\nu)$ in electronic recombinations to the states of principal quantum number s of the Z-ions is

$$(7\cdot 7) \qquad \frac{2^9\pi^5\epsilon^{10}n_e(Z+1)^4}{(6\pi)^{3/2}m_e^{1/2}h^2c^3s^3(kT)^{3/2}} \,.\, e^{(x-h\nu)/kT} \,.\, d\nu \; (h\nu > x)$$

per $(Z+1)$-ion per unit time, where x is the ionization potential of the state in question, and ϵ is the electronic charge. Dividing (7·7) by $h\nu$ and integrating with respect to ν from x/h to ∞ gives the recombination rate per $(Z+1)$-ion into this state. If we put $u = x/kT$, $y = h\nu/kT$ and remembering that

$$(7\cdot 8) \qquad x = 2\cdot 15 \times 10^{-11}(Z+1)^2/s^2 \text{ ergs}$$

for hydrogen-like ions, then the resulting integral can be expressed as

$$(7\cdot 9) \qquad 4\cdot 35 \times 10^{-25}\left(\frac{2}{m_e}\right)^{\frac{1}{2}} \frac{n_e(Z+1)^2}{sT^{\frac{1}{2}}} \,.\, ue^u \int_u^\infty \frac{e^{-y}}{y} \,.\, dy.$$

This expression may be used as giving a reasonable approximation for the recombination probability for the positive ions occurring in the corona, provided we interpret s as the principal quantum number of the outermost shell of the ion in question (the main contribution comes from the recombination to the ground state of the Z-ions). In all applications of (7·9) to recombinations to the ground state, u is sufficiently large for $ue^u \int_u^\infty \frac{e^{-y}}{y} dy$ to be replaced by unity without appreciable error. Then (7·6) and (7·9) enable (7·4) to be expressed in the form

$$(7\cdot 10) \qquad \frac{n_Z}{n_{Z+1}} = 1\cdot 7 \times 10^{13} \frac{x^{3/2}(Z+1)^2 e^{x/kT}}{lsT^{\frac{1}{2}}}.$$

The application to hydrogen is given by putting

$$x = 2\cdot 15 \times 10^{-11} \text{ ergs}, \; Z = 0, \; l = s = 1,$$

which gives the following values of $n_p/(n_p+n_n)$:

Table 9.—Ionization of hydrogen when radiation is absent

T in °K.	10,000	12,000	15,000	20,000	25,000
$n_p/(n_p+n_n)$	$8\cdot 8\times 10^{-3}$	$1\cdot 2\times 10^{-1}$	$6\cdot 7\times 10^{-1}$	$9\cdot 7\times 10^{-1}$	$9\cdot 99\times 10^{-1}$

In contrast with the thermodynamic case, the present values are independent of the electron density, but this result depends on an assumption tacitly made in the above work. Thus the ionization probability per free electron was taken as independent of n_e. This is valid in the chromosphere and corona, but for sufficiently high n_e the free electron states become densely packed for energies of order kT (on account of the exclusion principle), and this has the effect of appreciably reducing the ionization probability.

Remembering the different values of the temperature that must be used in (7·1) and (7·2) it is seen from Tables 8 and 9 that in the main body of the chromosphere (where $\log P_e$ is near

Table 10.—Ionization of neutral metal atoms under thermodynamic equilibrium

T \ $\log P_e$	6	4	2	0	−1	−2	−3
3600			$2·2\times10^{-5}$	$2·2\times10^{-3}$	$2·2\times10^{-2}$	$1·83\times10^{-1}$	$6·91\times10^{-1}$
3877		$1·7\times10^{-6}$	$1·7\times10^{-4}$	$1·7\times10^{-2}$	$1·43\times10^{-1}$	$6·25\times10^{-1}$	$9·42\times10^{-1}$
4200		$1·3\times10^{-5}$	$1·3\times10^{-3}$	$1·12\times10^{-1}$	$5·58\times10^{-1}$	$9·25\times10^{-1}$	$9·92\times10^{-1}$
4582		$9·5\times10^{-5}$	$9·5\times10^{-3}$	$4·88\times10^{-1}$	$9·06\times10^{-1}$	$9·90\times10^{-1}$	
5040	$7·4\times10^{-6}$	$7·4\times10^{-4}$	$7·1\times10^{-2}$	$8·82\times10^{-1}$	$9·86\times10^{-1}$		
5600	6×10^{-5}	6×10^{-3}	$3·76\times10^{-1}$	$9·83\times10^{-1}$	$9·98\times10^{-1}$		
6300	5×10^{-4}	$4·8\times10^{-2}$	$8·34\times10^{-1}$	$9·98\times10^{-1}$			
7200	$4·3\times10^{-3}$	$2·99\times10^{-1}$	$9·77\times10^{-1}$				
8400	$3·7\times10^{-2}$	$7·96\times10^{-1}$	$9·97\times10^{-1}$				

zero) and in the corona, collisional ionization of hydrogen is far more important than radiative ionization. This result applies *a fortiori* to ions with ionization potentials greater than 13·53 eV. On the other hand, radiation from the photosphere is important in the ionization of neutral metal atoms, which have ionization potentials lying between about 4 and 8 eV. Table 10 gives the fraction of atoms that are ionized when we put $x = 7·9$ eV. in the right-hand side of (7·3), and replace n_p, n_n by the corresponding densities for the metal atoms. This procedure ignores a slight correction that may occur in specific cases due to statistical weight factors. The values given in Table 11 show that only about 2% of atoms remain neutral when $\log P_e = 0$, $T = 5600°$ K. This somewhat exaggerates the degree of ionization since a thermodynamic calculation does not take account of the decrease of intensity within the Fraunhofer lines. The reduced emission in these lines is important because much of the interchange between the radiation and the atoms takes place at resonant wavelengths.

Finally, we may consider the application of (7·10) to the heavy positive ions occurring in the corona. In particular, results for *Fe* are given in Table 11.

Table 11.—*Ionization of iron when radiation is absent*

Z	s	$x(Z)$ in eV.	$T(n_Z/n_{Z+1} = 1)$ in °K.
1	4	16·5	$1{\cdot}5 \times 10^4$
2	4	30·5	$3{\cdot}1 \times 10^4$
3	4	56·8	$6{\cdot}6 \times 10^4$
4	4	~80	$\sim 10^5$
5	4	~100	$\sim 1{\cdot}4 \times 10^5$
6	4	~125	$\sim 2 \times 10^5$
7	4	150	$2{\cdot}8 \times 10^5$
8	3	233	$3{\cdot}7 \times 10^5$
9	3	261	$4{\cdot}4 \times 10^5$
10	3	289	$5{\cdot}2 \times 10^5$
11	3	325	$6{\cdot}3 \times 10^5$
12	3	355	$7{\cdot}4 \times 10^5$
13	3	390	$8{\cdot}9 \times 10^5$
14	3	454	$1{\cdot}2 \times 10^6$
15	3	487	$1{\cdot}4 \times 10^6$
16	2	1260	$3{\cdot}5 \times 10^6$

The quantity $T(n_Z/n_{Z+1} = 1)$ is the temperature, calculated from (7.10), that gives $n_Z/n_{Z+1} = 1$.

8. Intensities in the Chromosphere

(i) *The emission lines*

Chromospheric line intensities have been measured during total solar eclipses. The observations then give the apparent brightness per unit area of the chromosphere seen in projection against the sky.† By assuming (*a*) an exponential variation of density with height, (*b*) that the chromosphere is optically thin, Cillié and Menzel have converted this data into a rate of emission per unit volume, which is given as a function of height above the photosphere. But it now seems doubtful whether (*b*) can be applied near the base of the chromosphere in the case of strong lines such as *Hα*, *Hβ*, Hγ, *He* 5875·6, *Ca* I 4226·7, *H* and *K* of *Ca* II, *Mg* 3838·3, *Ti* II 3372·8, *Sc* II 3630·7, *Cr* II 3368·2. The observed data are not readily reduced to a useful form in these cases.

† C. R. Davidson and F. J. M. Stratton, *Mem. R.A.S.*, **64**, 105, 1927. A. Pannekock and M. Minnaert, *Verk. d. K. Akad. V. Wet. Ams.*, **13**, No. 5, 1928. C. R. Davidson, M. Minnaert, L. S. Ornstein, and F. J. M. Stratton, *M.N.*, **88**, 536, 1928. G. G. Cillié and D. H. Menzel, *Havard Circular*, No. 410. The latter authors apply the term ‘ base of the chromosphere ’ to a height ~600 km., while we use the term as applying to a height ~1500 km.

Although detailed radiation processes are best discussed by the methods given in section 20, we may note here, that if the chromosphere is optically thick in a given line, then a useful approximation to the rate of escape of radiation can be obtained from

$$(8\cdot 1)\qquad \frac{2\pi R^2 \nu_0{}^5 \Delta\lambda}{c^3(e^{h\nu_0/kT} - 1)} \text{ ergs/sec.,}$$

where ν_0 is the central frequency of the line, $\Delta\lambda$ is the effective width (in cm.) and T is the excitation temperature (as shown in section 20, T can be taken as $\sim 4830°$ K. even in the case of hydrogen). The formula (8·1) gives the total rate of radiation by the whole sun.

(ii) *The Balmer continuum*

The total observed intensity in the Balmer continuum is comparable with the total emission in all spectral lines together. Since the absorption cross-section for quanta in the continuum is very much less than the cross-section at resonant wavelengths, the chromosphere is certainly optically thin so far as radiation in the continuum is concerned. Thus the method of Cillié and Menzel can be used to obtain the rate of emission per unit volume from the observed intensities. For a height ~ 600 km. above the photosphere the emission per unit volume at λ 3640 A. is

$$\sim 3\cdot 8 \times 10^{-16} \text{ ergs/cm.}^3 \text{ per sec. per unit frequency range.}$$

The emission varies approximately with height h (km.) according to the factor $e^{-h/650}$.

The electron density can be obtained as a function of height by equating the rate of emission per unit volume, derived from observation, with the value calculated from (7·7). On multiplying (7·7) by n_p and putting $Z = 0$, $s = 2$, and x equal to ionization potential 3·38 eV. of the $s = 2$ states, we obtain the expression

$$(8\cdot 2)\qquad 2\cdot 63 \times 10^{-33} \frac{n_p n_e}{T^{3/2}} \cdot e^{(x-h\nu)/kT} \text{ per cm.}^3 \text{ per sec. per unit frequency,}$$

for the calculated emission. Since the assembly must be electrically neutral we must have $n_p = n_e$ to a close approximation. Moreover, at λ 3640 A. the factor $(x - h\nu)$ is zero and we therefore obtain

(8·3) $$n_e = 3{\cdot}8 \times 10^8 T^{3/4} e^{-(h-h_0)/1300} \text{ per cm.}^3,$$

where h_0 corresponds to a height of 600 km. above the photosphere. Table 12 gives n_e at various heights.

Table 12

Height above photosphere (km.)	600	1500	3500
T (°K.)	4830	4830	20,000
n_e (per cm.³)	$2{\cdot}2 \times 10^{11}$	$1{\cdot}1 \times 10^{11}$	$6{\cdot}6 \times 10^{10}$

These values of T are chosen in accordance with the results of chapters III and IV.

The value obtained for n_e at a height of 600 km. is confirmed by an independent estimate given by the equation

(8·4) $$\log c^* = 23{\cdot}26 - 7{\cdot}5 \log s^*,$$

obtained by Inglis and Teller † from a consideration of the termination of the Balmer series by the Stark effect. The quantity s^* is the principle quantum number of the last resolved line, and at temperatures $<10^5/s^*$, c^* is the density of positive ions *plus* n_e, while at temperatures $>10^5/s^*$, c^* is simply the positive ion density (the electrons making no contribution to the broadening because their velocities are too high). The value of s^* given by observation is 37, so that the second case must be applied. Thus since the contribution of multiply charged ions can be neglected on account of the great hydrogen abundance, we must have $n_e = c^*$ for an electrically neutral assembly. Hence we obtain $n_e = 3{\cdot}16 \times 10^{11}/\text{cm.}^3$ by putting $s^* = 37$ in (8·2).

9. Non-Steady Phenomena

(i) *The Corona*

When viewed directly the shape of the corona is found to change with the solar cycle. Although the precise relation between the phase of the sunspot period and the corona varies somewhat from one cycle to another, in general the corona is approximately circular at sunspot maximum. When the maximum is past the coronal streamers draw away from the poles and the longest rays are found in the sunspot zones. At sunspot minimum the typical

† D. R. Inglis and E. Teller, *Ap.J.*, **90**, 439, 1939.

corona possesses long equatorial streamers and short plume-like polar brushes. In particular, at the eclipse of 1878 equatorial streamers were detected out to about ten times the solar radius.

The change in the shape of the corona is not appreciable, however, at distances r from the solar centre $<1 \cdot 5$ (R as unit). Thus so far as the electron density is concerned this inner part of the corona is not affected by the solar cycle.† Ludendorff has shown that if contours of constant intensity are traced in a coronal photograph then the ratio of the polar and equatorial diameters of such contours vary with r according to the formula

$$a + b(1 - r), \qquad (9 \cdot 1)$$

where a is a coefficient close to 0·98 that is substantially independent of the solar cycle. The coefficient b on the other hand varies with the solar cycle and is approximately zero at spot maximum and decreases to about $-0 \cdot 2$ at spot minimum. Thus, whereas at sunspot minimum the ellipticity becomes important for $r > 2$, the corona is always nearly circular for $r < 1 \cdot 5$.

Although electron scattering remains approximately constant for $r < 1 \cdot 5$, there are important changes in the coronal emission lines within this region. The emission lines appear to follow a general cycle similar to that of the outer corona. The connexion between coronal emission and the outer corona is also shown by a fairly close correlation between individual bright-emission patches and particular streamers in the outer corona. Both these phenomena are connected with sunspots and with M-regions (see part (iv) of this section).

(ii) *Prominences*

The top of the chromosphere was defined in section 6 as the height ($\sim$12,000 km.) at which the $H\alpha$ line fades out. Prominences are exceptional regions, lying above this height, that are still visible in the ordinary lines of hydrogen, helium, calcium, strontium, titanium, magnesium, etc. A typical prominence has a height $\sim$35,000 km., but the heights vary widely from small extensions above the chromosphere up to very large values of $\sim$700,000 km.

It will be seen in chapters III and IV that the corona is normally

† H. A. Brück, *M.N.*, **104**, 33, 1944.

too hot to be visible in ordinary spectral lines. This means that prominences *must be regions of local cooling in the corona.*

The frequency and distribution of prominences are illustrated in Table 13, which refers to prominences with heights $>$20,000 km. In each solar hemisphere there are two zones of maximum frequency. One lies between latitudes 20° and 40° and follows the sunspot period, whereas the second zone lies in higher latitudes and is approximately half a cycle out of phase with the first zone. The prominences in the second zone break out between latitudes 50° and 60° two or three years after sunspot maximum, and thereafter the zone drifts towards the pole, which it reaches at the following maximum.

Prominences appear dark when observed (in the hydrogen lines, the *H*- and *K*-lines of calcium, etc.) in projection against the solar disk. For the line emission from a prominence is of the same order as, but less than, the emission by normal areas of the solar surface. The dark markings are often asymmetrically distributed about the centres of the lines. This asymmetry is attributed to the Doppler effect arising from the line of sight velocity of material in the prominence.

There is considerable variation in the shapes and properties of prominences. Pettit † has given a classification aimed at describing broad categories. In advancing this classification Pettit points out that there is no sharp distinction between one class and another. For example, prominences sometimes occur that show characteristics of two classes simultaneously.

Active prominences (class 1)

(1) When seen at the limb active prominences often have a tree-like appearance with a trunk rising from the chromosphere. Their general shapes change slowly and are retained for periods that range from a few hours up to several weeks.

(2) Streamers condensing in the inner corona flow into these prominences.

(3) Although they retain their general shape, knots of material are observed ‡ to condense and plunge down to the photosphere with velocities up to 100 km./sec. The trajectories followed by

† E. Pettit, *Ap.J.*, **76**, 9, 1932.
‡ R. R. McMath and E. Pettit, *Ap.J.*, **88**, 266, 1938.

this material enter the solar surface over an area with diameter as small as 10,000 km., and if prolonged below the surface the trajectories are found to intersect at a depth of about 10,000 km. below the photosphere. Thus these prominences are associated with centres of attraction on the solar surface, and accordingly they rotate with the sun. The centres of attraction do not require the presence of bright floculli.

(4) Active prominences may occur anywhere on the solar disk.

Eruptive prominences (class 2)

(1) Active prominences often develop sudden movements in approximately an outward radial direction. Such prominences are then referred to as eruptive.

(2) The average upward velocities are usually $\sim$100 km./sec., but velocities as low as 10 km./sec. occur, while at the other extreme a prominence has been observed by McMath and Pettit that had a maximum velocity of 728 km./sec. There is evidence for variations of velocity from one part of an eruptive prominence to another.

(3) Pettit has given the following rules governing the motion:

- (i) The velocity is uniform except that at intervals it increases suddenly.
- (ii) With few exceptions, the velocity following such a change is a small integral multiple of the value preceding the change.

The time required for a change of velocity may be as short as one minute.

(4) In extreme cases these prominences rise to heights $\sim R$ above the photosphere. As a prominence rises, matter pours down to the chromosphere along one or more trajectories, thereby often leading to an arched formation.

(5) As with active prominences the eruptive prominences are associated with centres of attraction on the solar surface, and they may occur anywhere on the disk.

Sunspot prominences (class 3a)

(1) These prominences show *streamers and bright jets* entering the spot zone over an area with diameter $\sim$50,000 km. The streamers make angles with the radial direction that are usually

less than 45°, and are distributed with approximate symmetry about this direction.

(2) The velocities along the streamers are of order 50 km./sec.

(3) The average height ~30,000 km. According to Pettit, coronal material condenses at this height and then flows down into the spot area.

(4) The streamers suggest that the direction of motion of material is controlled by the magnetic field of the associated sunspot group. The motion is independent of the magnetic polarity.†

Sunspot prominences (class 3b)

(1) Prominences of this subclass show *complex systems of arches and loops*. They are much rarer than prominences of class 3*a*.

(2) Their heights may be as great as 100,000 km., although smaller heights are more usual.

(3) According to McMath and Pettit a bright spot appears first in the corona. This spot then feeds the two arms of an arch. The condensed matter pours downwards with velocities that are often ~100 km./sec.

(4) This subclass is associated with complex spot groups. The motion of material again appears to be controlled by the magnetic field of the group, and is independent of magnetic polarities.‡

Sunspot prominences (class 3c)

(1) These prominences are associated with spot groups, but they *lie well outside the spot area*.

(2) Condensations move towards the spot area with velocities that range from 15 km./sec. up to ~100 km./sec. The velocity increases as the spot is approached.

(3) Condensations are attracted from distances as great as 250,000 km. The trajectories make much smaller angles with the horizontal than the streamers in subclass (*a*).

(4) These prominences occasionally rise bodily with high velocity and are torn apart. The remnants descend in thick streamers into a spot. A few hours later there is little trace left of the original prominence.§

† M. A. Ellison, *M.N.*, **104**, 22, 1944. ‡ M. Waldmeier, *Z.Ap.*, **14**, 91, 1937.
§ E. Pettit, *Ap.J.*, **84**, 334, 1936.

Sunspot prominences (class 3d)

(1) According to Newton † about 80% of solar flares of intensity 3 (see below) are accompanied by these prominences, which appear as narrow spikes moving in vertical trajectories, when viewed at the solar limb.

(2) They rise to a height ~30,000 km. and then fall back to the chromosphere. *The duration of this process is twenty minutes or less.*‡

The velocities are ~30 km./sec.

(3) The dark flocculi shown in projection against the solar disk have an excess of downward motions, but this is probably a selection effect.

Tornado prominences (class 4)

(1) These prominences have a spiral form resembling a screw. The motions are very rapid, and often the angular velocity becomes so high that the vortex explodes.

(2) They vary from ~5000 km. to ~20,000 km. in diameter, and from ~25,000 km. to ~100,000 km. in height.

(3) A faint smoke-like cloud issues from the top of the vortex and often bends over, in some cases touching the chromosphere.

(4) There is no appreciable horizontal drift.

Quiescent prominences (class 5)

(1) They appear at the limb as slowly changing dense clouds. This is an important difference from the types described above, which all have wisp-like structures that appear rather fragile when compared with the more solid appearance of the quiescent prominences. Cases are observed with lifetimes as long as two months.

(2) Material within these prominences moves with velocity ~20 km./sec. Thus 'quiescent' is employed as a relative term.

(3) The height is ~50,000 km. in many cases. When projected against the solar disk quiescent prominences appear as long dark filaments with widths ~10,000 km., and lengths that may exceed 200,000 km.

(4) They rotate with the sun and are associated with centres of attraction not connected with bright flocculi. Material pours down

† H. W. Newton, *M.N.*, **102**, 2, 1942. ‡ M. A. Ellison, *M.N.*, **102**, 11, 1942.

streamers extending between the prominence and the centre of attraction. The streamers are asymmetrically distributed with respect to the radial direction.

With the possible exception of prominences of class 3*d* it is probable that the general movement of the visible material in prominences is *downwards*. Even in eruptive prominences the apparent outward motion is not necessarily connected with the outward motion of visible material (indeed, material is observed to pour down to the solar surface), but rather with the rise of a *cooling influence*.

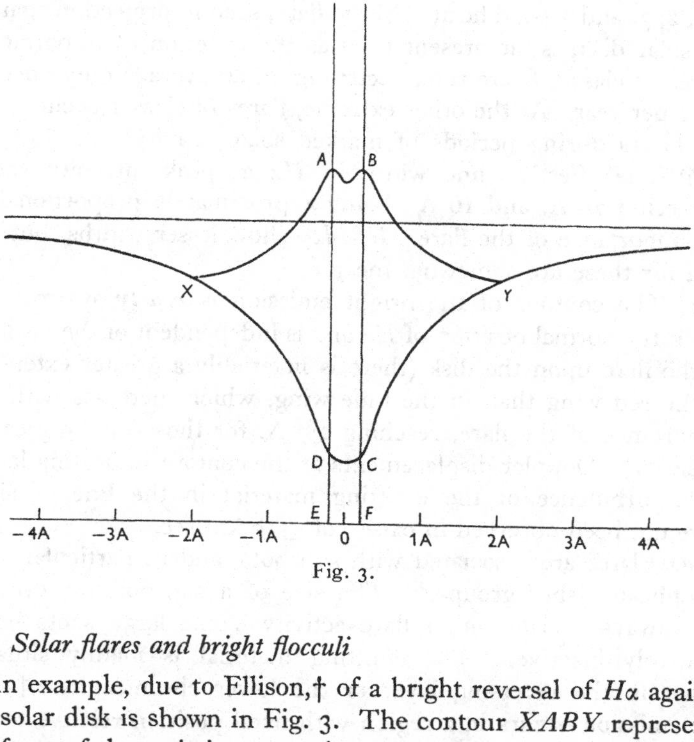

Fig. 3.

(iii) *Solar flares and bright flocculi*

An example, due to Ellison,† of a bright reversal of $H\alpha$ against the solar disk is shown in Fig. 3. The contour $XABY$ represents the form of the emission occurring over a limited area of the solar surface, whereas $XDCY$ is the normal $H\alpha$ contour of the surrounding background. When the slit of the spectroscope (width 0·45 A.) was placed centrally it covered a region between the lines

† M. A. Ellison, *M.N.*, **103**, 3, 1943.

AD and *BC*. The bright reversal is then observed with a maximum contrast that may be defined as $(c - 1)$, where c is the ratio of the areas *ABEF* and *DCEF*. The contrast is lost if the slit is moved from the centre to either of the points *X* or *Y*. We may describe the wavelength interval between these points as the *effective width* of the emission line. The emission also occurs in $H\beta$, $H\gamma$, . . . , in the lines of helium, in the *H*- and *K*-lines of *Ca* II, etc.

Flares are a particular class of bright reversal characterized by sudden commencements. The properties of flares are:

(*a*) They are roughly classified in order of increasing importance as 1, 2, 3, and 3+. The area of the flare, seen in projection against the solar disk, is, at present used as the criterion of importance. Flares of class 3+ are rare, occurring on an average only once or twice per year. At the other extreme, flares of class 1 occur every few hours during periods of marked solar activity.

(*b*) The effective line width in $H\alpha$ at peak intensity varies between 1·75 A. and 16 A., being approximately proportional to the importance of the flare. $H\beta$, $H\gamma$ show lesser widths, but the data for these are somewhat meagre.

(*c*) The contour of the bright emission is *nearly* symmetrical about the normal position of $H\alpha$ and is independent of the position of the flare upon the disk (there is invariably a greater extension in the red wing than in the blue wing, which increases with the importance of the flare, reaching 0·7 A. for those of the greatest intensity). Doppler displacements of the contour indicating large-scale turbulence of the emitting material in the line of sight have not been observed in excess of ± 10 km/sec.

(*d*) Flares are associated with sunspots, and in particular with complicated spot groups.† The size of a sunspot, however, is not always a criterion of flare activity, some large spots being relatively inactive. The emitting material is mainly situated either in the reversing layer or the lower chromosphere,‡ and the emission occurs in a region with fixed position relative to the position of the spot group. The areas of flares projected on the solar disk vary from a few hundred millionths up to values exceeding 10,000 millionths of the area of the disk. The duration

† H. W. Newton, *M.N.*, **103**, 244, 1943, and **104**, 4, 1944.
‡ R. G. Giovanelli, *Ap.J.*, **88**, 204, 1938.

of a flare is usually of the order of an hour or less, but lifetimes > 5 hours occasionally occur.

(*e*) Flares are strongly correlated with a number of terrestrial effects. Radio fade-outs, due to increased ionization in the *D*-layer, occur simultaneously with the visible appearance of intense flares. Great magnetic storms are associated † with flares of classes 3 and 3+. The magnetic disturbances commence about 26 hours after the appearance of the flare, and are most marked when the flare is near the centre of the disk. Finally, there is a growing body of evidence that the sun emits exceptionally high intensities in the radio metre wave-band during flares (see chapter VII).

The bright quiescent flocculi follow the solar cycle and occur mainly in the spot zones. They are usually observed in $H\alpha$ or the *H*- and *K*-lines of *Ca* II. Different levels in the chromosphere can be examined by admitting different line widths to the spectroscope. The central cores of the lines give the greatest heights, and reveal complicated structures in the neighbourhood of sunspots. If we exclude flocculi associated with prominences these structures are composed of small filaments and on occasion they show a pronounced vortical form. Richardson ‡ has recently analysed the sense of these vortices and his results confirm Hale's opinion that the phenomenon is of hydrodynamic origin. The region of the chromosphere overlying a spot acts as a sink, and on account of the solar rotation inflowing material tends to adopt a vortical structure characteristic of the hemisphere in which the spot occurs.

(iv) *M-regions*

In addition to the 'great' magnetic storms associated with particular solar flares, there are also disturbances of smaller intensity in the terrestrial magnetic field that show a 27-day periodicity. A suggestion made many years ago by Bartels is that these periodic disturbances can be ascribed to solar rotation if localized corpuscular-emitting areas of the solar surface are assumed to exist (the *M*-regions). An analysis of magnetic records shows

† W. M. H. Greaves and H. W. Newton, *M.N.*, **88**, 556, and **89**, 641.
‡ R. S. Richardson, *Ap.J.*, **93**, 24, 1941.

that, on this hypothesis, *M*-regions sometimes persist for as long as six months.

There has been a long search for visible features indicating the presence of *M*-regions. It is only recently, however, that this project has been successful, for there is no connexion with the more obvious features such as spots, bright flocculi, etc. In 1939 Waldemeier † proposed that regions of bright coronal line emission (the lines 5303 A. and 6374 A. being used) are identical with *M*-regions. This suggestion is supported by Allen,‡ and also by a detailed analysis due to Shapley and Roberts.§ The work of the latter authors leads to the conclusion that the probability of a terrestrial magnetic disturbance is considerably greater three days after the appearance of a region of bright coronal emission at the eastern limb of the sun, than at the time of meridian passage. This result suggests, in contrast with the emission from solar flares, that the main corpuscular emission in an *M*-region is sideways at angles greater than 45° with the radial direction. It may be noted that the work of Shapley and Roberts is mainly confined to the effect of bright coronal emission regions occurring on the eastern half of the sun. To establish fully their conclusions a similar analysis for the western half is required.

Allen has suggested 3 days as the average time of transit between the sun and the earth of the particles emitted by *M*-regions. There is support for this value in the work of Shapley and Roberts.

† M. Waldemeier, *Z.f.Ap.*, **19**, 21, 1939.
‡ C. W. Allen, *M.N.*, **104**, 13, 1944.
§ A. H. Shapley and W. O. Roberts, *Ap.J.*, **103**, 257, 1946.

CHAPTER III

THE CHROMOSPHERE AND CORONA: THEORY PART I

10. Accretion of Interstellar Material

It was suggested by Milne † that *Ca* II in the chromosphere may be supported against solar gravity by radiation pressure in the *H*- and *K*-lines. The theory became untenable, however, when it was realized that the calcium content of the solar atmosphere is extremely small compared with the hydrogen abundance. A thorough discussion of the problem by McCrea ‡ showed that radiation pressure could not supply more than one tenth of the upward force required to explain the existence of the hydrogen chromosphere. Although radiation pressure is consequently of little importance in the present discussion, it is of interest that the process suggested by Milne provides an explanation of the expulsion of calcium ions during solar flares. Following a suggestion due to Chapman, calcium ions travelling from the sun to the earth have been detected by Richardson and also by Brück and Rutllant.§ We shall return to this question in chapter v.

The very high kinetic temperatures now known to exist in the corona provide further evidence that radiation pressure does not play the dominant role in the solar atmosphere. It seems clear that *mass motions* of material with velocities at least of the same order as the thermal velocities of the protons in the corona are required. Now the mean thermal velocity of a proton at $2 \cdot 10^6$ deg. K. is 205 km./sec. This is of an order of magnitude that suggests a connexion with the velocity $\sim$600 km./sec. with which a particle falling freely from rest at infinity would enter the sun. Thus we are led to consider the effect of the capture of interstellar material, since accreted material falls into the sun with a velocity

† E. A. Milne, *M.N.*, **84**, 354, 1924; **85**, 111, 1924; **86**, 8 and 578, 1925–26.

‡ W. H. McCrea, *M.N.*, **89**, 483, 1928–29.

§ R. S. Richardson, *Ann. Report Mt. Wilson Obs.*, 1943–44. H. A. Brück and F. Rutllant, *M.N.*, **106**, 130, 1946.

of the required order.† The problem of the capture of interstellar material has recently been considered in detail, but an earlier discussion is sufficient for the present purpose.‡

Imagine interstellar material to be streaming past the sun, from right to left in Fig. 4, and let the velocity of any element of it relative to the sun when at great distances be v. Consider the part of the cloud that if undeflected by the sun would pass within

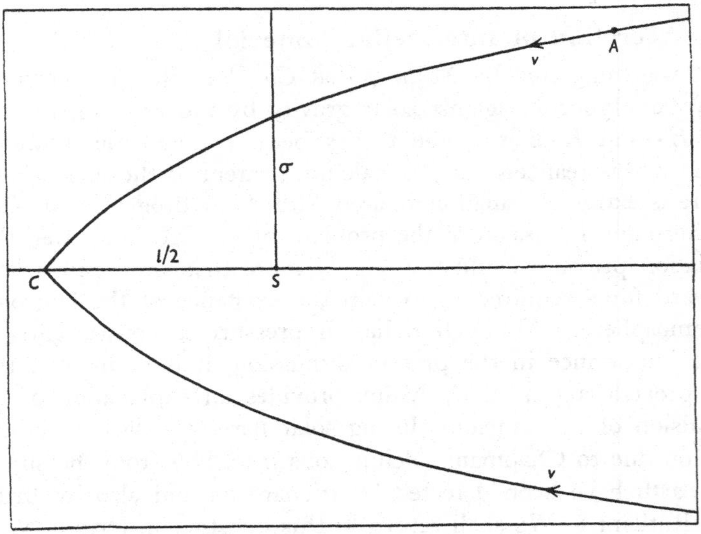

Fig. 4.

a distance σ or less of its centre. It is clear that collisions will occur to the left of the sun because the attraction of the latter produces two opposing streams of particles that tend to meet on the line CS. The effect of such collisions is to destroy the angular momentum about the sun of the individual elements of material. If after collision the surviving radial component of the velocity is insufficient to enable the particles to escape, then, provided pressure effects are neglected, such particles will eventually be swept into the sun. Suppose, for example, that an element of

† It has been suggested by V. Vand, *Nature*, **151**, 728, 1943, that the temperatures in the corona may arise from heating due to the infall of *interplanetary* material, but it is doubtful whether interplanetary material could be drawn into the sun at the rate necessary to explain the coronal temperatures.

‡ F. Hoyle and R. A. Lyttleton, *Proc. Camb. Phil. Soc.*, **35**, 405, 1939. H. Bondi and F. Hoyle, *M.N.*, **104**, 273, 1944.

volume of the cloud A, whose initial angular momentum is $v\sigma$, loses this momentum through its constituent particles suffering collisions near C, then the effective capture radius σ can be calculated such that the velocity radially at C is equal to the escape velocity at this distance. The element describes a hyperbola whose equation, with the usual notation, is

$$\frac{l}{r} = 1 + e \cos \theta. \tag{10·1}$$

The direction parallel to the initial asymptote corresponds to $(e \cos \theta + 1) = 0$, and hence the direction SC corresponds to $e \cos \theta = 1$. Thus the distance SC is simply $l/2$. Since $r^2 d\theta/dt = h$, and is constant, the radial component of velocity is given by

$$dr/dt = eh \sin \theta / l. \tag{10·2}$$

Now $h = (\mu l)^{\frac{1}{2}} = v\sigma$, where μ/r^2 is the attraction of the sun at distance r. Hence, at C, dr/dt has the value v, the velocity of the cloud at infinity. The part of the cloud will accordingly fail to escape after collision if

$$v^2 = 2\mu/SC = 4\mu/l = 4\mu^2/v^2\sigma^2. \tag{10·3}$$

Accordingly, the effective capture radius σ is $2\mu/v^2$. If G is the constant of gravitation and M is the mass of the sun, then $\sigma = 2MG/v^2$. For $v = 20$ km./sec., σ is as large as 1000 R, which is comparable with the radius of Jupiter's orbit. The total number of particles reaching the sun/sec. is simply the number crossing an area bounded by a circle of radius $2GM/v^2$ perpendicular to the direction of v, the velocity of the cloud. If τ_∞ is the space density of interstellar hydrogen atoms at great distances from the sun, then this number is

$$4\pi G^2 M^2 \tau_\infty / v^3. \tag{10·4}$$

The captured material forms a condensation between C and S, and falls towards S. At first sight it seems that the material enters the sun along CS. But it is easy to see that this is not the case. Consider for the moment a single particle falling from C to S. This particle will enter the sun along CS only if it has zero velocity transverse to CS. A very small transverse velocity of order Rv/σ, $(R/\sigma \sim 10^{-3})$, when the particle is at C is sufficient to produce a marked alteration in the direction from which the particle enters the sun, and a transverse velocity appreciably greater than Rv/σ results in the particle not entering the sun at all, but in returning approximately along SC in a highly elliptic orbit. Returning now to the condensation of interstellar gas between C and S, we notice that although the thermal motions in this gas must be less than v on account of the requirement that pressure effects do not appreciably affect the rate of capture of interstellar material, the thermal velocities can nevertheless be *of order* v, thereby exceeding Rv/σ by a factor of order 10^3. Accordingly, the material readily spreads round the sun as it moves inwards. For σ comparable with the radius of Jupiter's orbit, it is considered that the captured material becomes approximately isotropic with respect to the sun when it has moved inside the Earth's orbit.

In the discussion of the following sections we shall idealize the way in which material arrives at the solar surface by assuming strict isotropy and we also take the velocity V of the captured material near the sun to be everywhere radial, and to be given by $(2GM/r)^{\frac{1}{2}}$, (r in cm.). These assumptions are made in order to simplify a rather difficult hydrodynamic problem.

If τ is the space density of captured hydrogen atoms at the sun, then according to these assumptions the rate at which atoms are captured is $4\pi R^2 V\tau$.

This must be equal to (10·4), so that we have

$$\tau_\infty = 4R^2 V\tau/\sigma^2 v. \qquad (10\cdot5)$$

11. Region 2†

The incoming atoms of accreted hydrogen are stopped by collisions with the material of the solar atmosphere. The region of penetration of the incoming atoms will be referred to as region 2, and its base is defined as the level at which the average downward momentum has been reduced by collisions to a fraction $1/e$ of the original value. This level is shown later to be $\sim 1\cdot3 \times 10^{10}$ cm. above the photosphere. There are two stream lines through each point of region 2; one stream being due to the incoming material and the other to the motion of the solar atmosphere.

We now estimate the amount of material belonging to the solar

† The remainder of this chapter is taken from H. Bondi, F. Hoyle, and R. A. Lyttleton, *M.N.*, **107**, 184, 1947.

atmosphere that is present in region 2. In this calculation we anticipate the result obtained in chapter IV that the incoming hydrogen is ionized and, on account of the overwhelming abundance of hydrogen, we neglect the effect of collisions with other elements. Furthermore, the temperature later calculated for region 2 is so high that the hydrogen in this region can be regarded as wholly ionized. Hence the collisions of importance in stopping the accreted hydrogen are proton-proton collisions.

The number of hydrogen atoms of the solar atmosphere, lying in a unit radial column (cross-section 1 cm.^{2}) with base at distance r from the solar centre is

(II·1) $$\int_r^\infty n_p(y)dy = N(r), \quad \text{say,}$$

where n_p is the density of the protons belonging to the solar atmosphere. Now the value of $N(r)$, $N(r^*)$, say, for which the average downward momentum of the incoming hydrogen atoms is reduced to a fraction $1/e$ of its original value is given by

(II·2)
$$1/N(r^*) = \text{collision cross-section in region 2 between incoming protons and protons of the solar atmosphere.}$$

This gives $N(r^*)$ but not r^*, which is later shown to be $\sim 8{\cdot}2 \times 10^{10}$ cm.

The probability of scattering through an angle lying between θ and $(\theta + d\theta)$ in a collision of relative velocity V between protons is given in the usual notation by †

(II·3)
$$2\pi I(\theta) \sin\theta\, d\theta = 8\pi\left(\frac{\epsilon^2}{m_p V^2}\right)^2 \left[\sec^4\theta + \operatorname{cosec}^4\theta + 2\cos\left(\frac{2\pi\epsilon^2}{hV}\log\tan^2\theta\right)\sec^2\theta\operatorname{cosec}^2\theta\right]\sin\theta\cos\theta\, d\theta,$$

where m_p is the mass of the proton. The average loss of momentum in a proton-proton collision is therefore given by

(II·4) $$2\pi m_p V \int (1 - \cos^2\theta) I(\theta) \sin\theta\, d\theta.$$

A difficulty, due to the divergence of this integral for small θ,

† N. F. Mott and H. S. W. Massey, *The Theory of Atomic Collisions*, Oxford, 1933, p. 73.

is overcome by cutting off at lower a limit $\theta_{\min}$, say. Then (11·4) leads to the collision cross-section

$$(11\cdot5) \qquad 8\pi(\epsilon^2/m_p V^2)^2 \ln(1/\theta_{\min}).$$

The lower limit $\theta_{\min}$ has a definite physical significance, which can be seen by resorting to a classical picture (this is justified because only a very small contribution arises from the term in (11·3) that contains h). Let ξ be the closest distance of approach between the two protons. Then, for collisions in which the deviation θ of the direction of motion of the protons is small, it can be shown that

$$(11\cdot6) \qquad \theta = 2\epsilon^2/\xi m_p V^2.$$

Accordingly, a cut-off for θ requires an upper limit to be taken for ξ. Now such an upper limit for ξ does exist, corresponding approximately to the mean distance between the particles in region 2. This cut-off distance is about 10^{-3} cm. for the densities inferred below. Thus $\theta_{\min} \sim 2 \times 10^3 \, (\epsilon^2/m_p V^2)$. Although the value of $\theta_{\min}$ determined in this way can only be regarded as an order of magnitude estimate, the determination of the cross-section (11·5) is not appreciably affected by the uncertainty, since $\theta_{\min}$ appears only in a logarithmic term. Thus from (11·2) and (11·6) it follows that

$$(11\cdot7) \qquad 1/N(r^*) = -8\pi(\epsilon^2/m_p V^2)^2 \ln\{2 \times 10^3(\epsilon^2/m_p V^2)\},$$

which gives $N(r^*) = 1\cdot36 \times 10^{18}$ atoms/cm.^{2} when we put

$$V = (2GM/r^*)^{\frac{1}{2}} = 570 \text{ km./sec.}$$

(we anticipate the result $r^* = 8\cdot2 \times 10^{10}$ cm. given by later work).

The momentum $\tau m_p V^2$ per cm.^{2} of the incoming hydrogen atoms is distributed among the $N(r^*)$ particles per cm.^{2} (we neglect the penetration of the incoming atoms beyond the base of region 2), producing an average force $\tau m_p V^2/N(r^*)$ acting on these particles in the same direction as solar gravity. Thus the average downward acceleration of an atom of hydrogen belonging to the solar atmosphere $\sim g(1+q)$, where $q = \tau V^2/gN(r^*)$ and $g = MG/(r^*)^2$ is the value of gravity at distance r^* from the solar centre. The parameter q, which depends on the accretion rate, is shown later to be $\sim 0\cdot25$. Thus, since τ can be expressed in the form

(11·8) $\tau = N(r^*)q/2r^* = 8\cdot 3 \times 10^6 q$ atoms per cm.^{3},

we obtain $\tau \sim 2\cdot 1 \times 10^6$ atoms/cm.^{3} Then by putting $\sigma = 2MG/v^2$, $v = 10$ km./sec. in (10·5) we have $\tau_\infty \sim 44\cdot 8$ atoms/cm.^{3}, which corresponds to a mass density $\sim 7\cdot 43 \times 10^{-23}$ gr./cm.^{3} This value is well within the range given by studies of galactic structure and stellar evolution.

If the material in region 2 possesses an average kinetic temperature Θ, then the distribution of material in this region is given approximately by the hydrostatic formula

(11·9) $n_p(r) = n_e(r) = n_p(r^*)e^{-(r-r^*)/H}, \; r > r^*,$

where $H = \mathfrak{R}\Theta/g\mu(1+q)$, $\mathfrak{R}$ is the gas content, and μ is the mean molecular weight in region 2. From (11·1) and (11·9) we have

(11·10) $n_p(r^*) = N(r^*)/H.$

Finally, we note that incoming electrons are more readily stopped by collisions than the incoming protons. Thus in order to preserve electrical neutrality in region 2 there must be an electric field that produces a downward drift of electrons. This question has been investigated and it has been found that the required electric field does not have an appreciable effect on the motion of the incoming protons. Thus our calculation of $N(r^*)$ is substantially unaffected by this field.

12. The Temperature in Region 2

The rate at which energy is added to region 2 by the incoming hydrogen atoms is $\tau m_p V^{3/2}$ per unit area of the solar surface. Remembering that $q = \tau V^2/gN(r^*)$, this expression becomes $9\cdot 4 \times 10^{-13} qN(r^*)$ ergs/sec. per cm.^{2} The processes available for disposing of this energy are radiation, conduction, and convection. We consider these processes in turn and it will be shown that convection is by far the most important.

(i) *Radiation*

The contribution to the rate of radiation per unit volume by the free-bound transitions of hydrogen at temperature Θ is obtained from (7·7) by putting

$$T = \Theta, \quad Z = 0, \quad x = 13\cdot 53(1 - 1/s^2)\,\text{eV.},$$

multiplying by n_p, integrating with respect to ν from x/h to ∞, and finally summing with respect to s from 1 to ∞. This has been carried out by Minkowski † who also gives the contribution of the free-free transitions. The result is

$$(12{\cdot}1) \quad 1{\cdot}45 \times 10^{-27}\,\Theta^{\frac{1}{2}}[1 + 3{\cdot}85 \times 10^5/\Theta]n_e n_p \text{ ergs/cm.}^3\text{/sec.},$$

where Θ is in deg. K. This expression, together with (11·9), (11·10), and the condition $n_e = n_p$ of electrical neutrality, gives

$$(12{\cdot}2) \quad 7{\cdot}25 \times 10^{-28} N^2(r^*) g\mu(1 + q)\mathfrak{R}^{-1}\Theta^{-\frac{1}{2}}[1 + 3{\cdot}85 \times 10^5/\Theta]$$

for the total radiation per sec. by the hydrogen lying in a unit radial column (cross-section 1 cm.^{2}) with base at the base of region 2.

The maximum rate of radiation occurs for the lowest value of Θ consistent with the hydrogen being effectively wholly ionized (the use of (11·9) and (11·10) assumes the hydrogen to be ionized). According to Table 9 this condition corresponds to $\Theta \sim 20{,}000$ deg. K. With this value, and with $N(r^*) = 1{\cdot}36 \times 10^{18}$ atoms/cm.^{3}, $\mu = {\cdot}5$, we obtain

$$(12{\cdot}3) \qquad 1{\cdot}6 \times 10^{-14}(1 + q)N(r^*) \text{ ergs/cm.}^2 \text{ per sec.}$$

for the maximum rate of radiation. This expression is too small by a factor of more than ten to offset the rate $9{\cdot}4 \times 10^{-13} qN(r^*)$ ergs/sec. per cm.^{2} at which the incoming accreted hydrogen adds energy to region 2 (it is shown below that $q \sim {\cdot}25$).

On account of their low abundance metallic atoms have a radiation rate per unit volume that is small compared with hydrogen, unless the value of Z in (7·7) is of order 10. But such a value of Z requires Θ to be $\sim 10^6$ deg. K., and the emission by hydrogen at this high temperature is small compared with (12·3). Thus the rate of radiation by metallic atoms is in any case small compared with (12·3). Accordingly, we conclude that, in steady state conditions, radiation is not important in disposing of the energy supplied to region 2 by the incoming accreted material.

(ii) *Conduction*

The temperature gradient necessary for the energy of the accreted material to be transported downwards by conduction

† R. Minkowski, *Ap. J.*, **96**, 199, 1942.

can be obtained from classical theory.† In the present problem the number of particles per cm.^{3} multiplied by the mean free path is $\sim 10^4 \Theta^2$ cm.$^{-2}$. Using this value it is found that conduction cannot be important unless the temperature rises from 10,000° K. to 10^5 deg. K. in a range of height < 10 km. Such a variation of temperature would lead to an enormous outward pressure gradient and could not represent a static arrangement of material.

(iii) *Convection*

We first show that region 2 is convectively unstable. Convection currents will only arise if every static arrangement of material is unstable. Consider a particular small element of gas and let this material be moving downwards while neighbouring material remains static. Owing to its motion the relative velocity of the incoming accreted material is diminished and therefore by (11·5) more of the accreted material is absorbed than would occur in the static case. Accordingly, the element of volume in question receives more downward momentum than its static neighbours. Similarly, it can be seen from (11·5) that gas moving up would receive less downward momentum than its static neighbours. Thus the absorption of incoming material contributes a factor of instability that will tend to produce convection. This tendency becomes strong when q is of order unity, since convection currents would be driven by a force of the same order as gravity itself.

In addition, it is also important that in a system of powerfully driven convection currents the downward columns be accelerated *as a whole* by receiving more downward momentum per unit cross-sectional area than their upward moving neighbours. It is easy to see that this condition cannot be satisfied if the columns move strictly vertically, for in this case the downward momentum per unit cross-sectional area supplied by accretion will on the average be the same for both ascending and descending columns. If, however, the columns are slanted so as to move at an angle to the radial direction, then the total downward momentum received per unit cross-sectional area by the descending columns can substantially exceed that for the ascending columns. For owing to excess absorption of accreted material in a slanting descending column the gas on its lower side will be shielded from

† G. Joos, *Theoretical Physics*, Blackie, 1940, p. 537.

the accreted material and must therefore receive less downward momentum than in the case where the columns are vertical. Accordingly this gas has a tendency to rise. An opposite state of affairs occurs for a slanting ascending column, which absorbs particularly little accreted material. Thus the gas on the lower side of a slanting ascending column must receive an exceptionally large share of accreted material and will hence be driven downwards. It is seen therefore that an alternating arrangement of parallel slanting ascending and descending currents must be powerfully driven by the incoming accreted material.

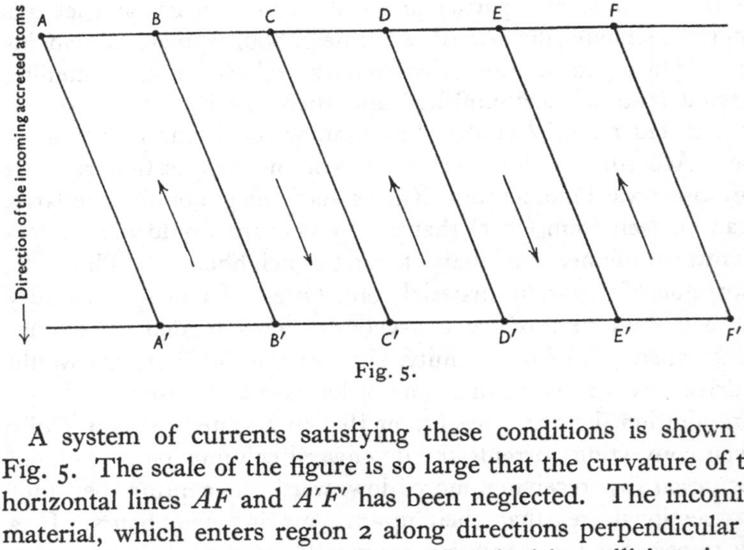

Fig. 5.

A system of currents satisfying these conditions is shown in Fig. 5. The scale of the figure is so large that the curvature of the horizontal lines *AF* and *A′F′* has been neglected. The incoming material, which enters region 2 along directions perpendicular to *AF*, is regarded as being effectively absorbed by collisions in the range of height between *A′F′* and *AF*. The material in the regions *ABB′A′*, *CDD′C′*, and *EFF′E′* is ascending, while the material in *BCC′B′* and *DEE′D′* is descending. For this arrangement of currents the absorption in *ABA′*, *CDC′*, and *EFE′* must be small compared with that in *BCB′* and *DED′*. Similarly, absorption in *BA′B′*, *DC′D′*, and *FE′F′* is small compared with that in *CB′C′* and *ED′E′*. This differential absorption drives the currents in the manner shown.

It is emphasized that the object of this discussion is to show that a system of driven currents can exist. It is not claimed that the system shown in Fig. 5 is the only possibility. Indeed, other systems may be even more powerfully driven than the present example. Thus the above argument shows that region 2 is unstable, but it does not define the exact system of currents that will occur. Judging by the complexity of convection problems the motion of material in region 2 is likely to be intricate.

We now make a preliminary estimate of the quantitative possibilities of convection. The most rapid way in which energy can be removed from region 2 is through convection currents moving with the velocity of sound in the solar material. If half the material in region 2 is regarded as moving downwards with velocity $(10\mathfrak{R}\Theta/3)^{\frac{1}{2}}$, this being the velocity of sound in ionized hydrogen at temperature Θ, it can be shown (see below) that the rate at which energy is convected out of region 2 is of order

$$(12\cdot4) \qquad 0\cdot5 m_p n_p(r^*)(10\mathfrak{R}\Theta/3)^{\frac{1}{2}}(5\mathfrak{R}\Theta/3 + 5\mathfrak{R}\Theta) \text{ per cm.}^2,$$

where the first term in the last bracket arises from the kinetic energy of mass motion of the material, and the second term is due to the thermal energy of the downward moving material. To offset the removal of material from region 2 by the downward convection currents there must be upward currents carrying material into region 2 from below. These upward currents will also carry thermal and kinetic energy. Thus the mean downward flux of energy is given by (12·4) only if the energy carried by the upward currents can be neglected in comparison with the energy carried by the downward currents. A full discussion of this question is given later. For the present, (12·4) will be accepted as giving an order of magnitude estimate for the energy removed from region 2. An estimate for Θ can then be obtained by equating (12·4) to the rate at which energy is added to region 2 by the accreted hydrogen. On substituting for $n_p(r^*)$ from (11·10) and writing $\mathfrak{R} = 8\cdot26 \times 10^7$, $\mu = \cdot5$, we obtain after a short calculation

$$(12\cdot5) \qquad \Theta \sim 10^6 q^2/(1+q)^2 \ {}^\circ\text{K}.$$

Thus for q of the order of unity Θ is of order 10^6 deg. K. Moreover, from the way Θ has been calculated it follows that (12·5) must give a lower limit to the kinetic temperature in region 2.

13. Region 1

This is a transition region possessing the reversing layer as lower boundary and the base of region 2 as upper boundary. This region has the remarkable property that while the pressure *decreases* outwards the temperature *increases* from less than 6000° K. at the lower boundary up to $\sim 10^6$ deg. K. at the upper boundary. Our present object is to show how this temperature variation is achieved. It is permissible in the following discussion to neglect radiation, for as shown in chapter IV, radiation is in general very small except in material immediately overlying the reversing layer. We also neglect the penetration of accreted atoms below the base of region 2. To this approximation, region 1 is a single stream region.

Region 2 is the source of powerful convection currents that carry downwards the energy supplied by accretion. Below region 2 the currents enter a region in which they are no longer driven since effectively all the accreted hydrogen is absorbed in region 2. Accordingly, in region 1 the downward currents continue only by virtue of the momentum they have acquired in region 2. Indeed it turns out that owing to the outward temperature gradient being positive the currents are gradually slowed down. The slowing-down process has the effect of building up a pressure at the bottom of region 1 that drives sufficient material upwards to compensate for the material transported by the downward currents. The downward currents are hotter than the ascending currents, and when radiation is neglected (this is valid except near the bottom of region 1) the difference between the energies carried by the downward and the upward columns must be equal to the energy supplied by accretion.

The present investigation differs from the usual problems of convection in that we have here a stable atmosphere (temperature increasing with height) into which powerful currents are projected from above. It is possible to effect progress with this rather formidable problem by setting up a number of bulk equations, but before considering these equations it is desirable to consider the following qualitative features:

(*a*) The motion of material in region 1 is regarded as mainly vertical. That is, the magnitude of the vertical pressure gradient is taken to be large compared with the horizontal pressure gradient. To a sufficient approximation the pressure is the same function of height above the photosphere in both ascending and descending columns.

(*b*) Conduction of heat from the descending to the ascending columns will be neglected. Conduction must be negligible unless the columns are very thin, and even in this case conduction would be small compared with the effect of direct interchange of material between the columns. It is well known that such a mixing process becomes strong when the columns are thin.

(*c*) The descending currents not only carry an excess of energy over the ascending columns, but also an excess of material. Thus just as the excess energy compensates for the energy supplied by accretion so the excess material compensates for the material added by accretion. In the work of sections 14 and 15 we neglect this downward transport of material. The results obtained on this basis, although always valid in the main body of region 1, must be treated with caution near the top of region 1. This question is considered in detail in section 16.

(*d*) The bulk equations derived in the following section refer to the averaged properties of the ascending and descending columns when taken as a whole. It is possible that individual currents may possess a motion that differs somewhat from these average equations.

(*e*) Solar gravity g will be taken, in region 1, as a constant independent of height.

(*f*) A positive outward temperature gradient and a negative outward pressure gradient cannot occur together unless there is an interchange of material between the ascending and descending columns, since otherwise there would be direct conflict with the adiabatic properties of a gas. Thus interchange of material between the columns must play a vital part in the following discussion.

(*g*) The surfaces of separation of the ascending and descending columns are defined in the following way. Take two points at the same height, one lying in an ascending column and the other in an adjacent descending column. Then in passing along the horizontal line joining the points the velocity of material changes from being strongly upward to being strongly downward. The point of zero vertical velocity on this line is defined as lying on the surface of separation of the two columns. Thus the streamlines are horizontal on the surfaces separating the columns.

14. The Equations of Motion

Let h be a height co-ordinate whose origin is as yet unspecified (the work of chapter IV shows that the origin is several hundred km. below the base of the chromosphere) and take ρ, T for the density and temperature in the descending columns. Similarly, let $\boldsymbol{\rho}$, $\boldsymbol{T}$ represent the density and temperature in the ascending columns.† We now take an *average* pair of columns, one ascending and the other descending. The cross-section of the downward column will be written as $S(h)$ and that of the ascending column as $\boldsymbol{S}(h)$. If $P(h)$ is the pressure at height h, then, by (a) of section 13, P is the same function of h in both columns. Moreover, if $Q(h)$ is the mass of downward material crossing $S(h)$ per unit time, then by (c) of section 13 the rate at which upward moving material crosses $\boldsymbol{S}(h)$ is also $Q(h)$. The amount of material crossing per unit range of height per unit time from the downward column to the upward column will be written as $U(h)$, and the transfer from the upward to the downward column will be denoted by $\boldsymbol{U}(h)$ per unit height range per unit time. To complete this set of definitions we denote the velocity in the downward column by $w(h)$ and the velocity in the ascending column by $\boldsymbol{w}(h)$.

These definitions lead to the following equations:

(14·1) $$Q = S\rho w = \boldsymbol{S}\boldsymbol{\rho}\boldsymbol{w},$$

(14·2) $$dQ/dh = U - \boldsymbol{U},$$

(14·3) $$P = 2\mathfrak{R}\rho T = 2\mathfrak{R}\boldsymbol{\rho}\boldsymbol{T},$$

since the mean molecular weight of the material is very close to 0·5. The derivation of further equations requires a knowledge of how material is transferred between the two columns. The rule adopted in subsequent work is:

Material is transferred from one column to another with negligible vertical velocity (*see* (g) *of section* 13), *but with the temperature of the column of its origin.*

† In the present chapter symbols in Clarendon type do not denote vectors.

(a) The energy equations

The energy of any element of volume consists of three parts: kinetic, potential, and thermal energy. The kinetic and potential energies per unit mass in a descending column are $0{\cdot}5w^2$ and gh respectively, while the thermal energy per unit mass is $3\mathfrak{R}T/2\mu$ or $3\mathfrak{R}T$ for $\mu = 0{\cdot}5$. Consider the region of a descending column lying in the range of height between h and $(h + \delta h)$. The energy flowing *out* per unit time across the plane faces of this region is

$$-\delta h \, . \, \frac{d}{dh}\left\{Q(0{\cdot}5w^2 + 3\mathfrak{R}T + gh)\right\},$$

and this energy must be equal to the sum of the following three parts:

(1) The rate at which work is done by the pressure on the material lying within the region in question. This rate is $\delta h \, . \, d(SPw)/dh$ which on using (14·1) and (14·3) becomes $2\mathfrak{R}\delta h \, . \, d(QT)/dh$.

(2) The rate at which energy is added across the curved surface of the region due to material transferred from the upward moving columns. From the definition of $\boldsymbol{U}$ together with the rule of mixing this is

$$\delta h \, \boldsymbol{U}(3\mathfrak{R}\boldsymbol{T} + gh) + 2\delta h\mathfrak{R}\,\boldsymbol{UT} = \delta h \, \boldsymbol{U}(5\mathfrak{R}\boldsymbol{T} + gh).$$

The second term on the left-hand side arises for reasons similar to the pressure term in (1), and is due to the work done by the pressure in moving the quantity of material $\delta h \, \boldsymbol{U}$ out of the ascending column.

(3) The rate at which energy is added to the region due to the transfer of material to the ascending columns. From the definition of u and the rule of mixing this is

$$-\delta h U(3\mathfrak{R}T + gh) - 2\delta h\mathfrak{R}UT = -\delta h U(5\mathfrak{R}T + gh),$$

where the second term on the left-hand side is due to the work done against the pressure in the transport of the quantity of material $U\delta h$ out of the descending column.

On collecting the various terms we obtain

(14·4)

$$\frac{d}{dh}\{Q(0{\cdot}5w^2 + 5\mathfrak{R}T + gh)\} = \boldsymbol{U}(5\mathfrak{R}\boldsymbol{T} + gh) - U(5\mathfrak{R}T + gh).$$

A similar equation can be obtained for the ascending column and is

$$(14{\cdot}5)\quad \frac{d}{dh}\{Q(0{\cdot}5\mathbf{w}^2 + 5\mathfrak{R}\mathbf{T} + gh)\} = U(5\mathfrak{R}T + gh) - \mathbf{U}(5\mathfrak{R}\mathbf{T} + gh).$$

Equations (14·4) and (14·5) determine the transport of energy. According to these equations

$$(14{\cdot}6)\quad Q\{(0{\cdot}5w^2 + 5\mathfrak{R}T) - (0{\cdot}5\mathbf{w}^2 + 5\mathfrak{R}\mathbf{T})\} = \text{constant},$$

which expresses the condition that the excess energy carried by the downward currents is a constant independent of height. Thus (14·4) and (14·5) satisfy the important requirement of constant downward energy transport (radiation being neglected). This pair of equations also satisfies the formal requirement that they must remain unchanged as a pair if the signs of g and h are changed and symbols in heavy type are interchanged with the symbols in ordinary type.

(b) The momentum equations

Consider again the region of a descending column lying in the height range h to $(h + \delta h)$. According to the rule of mixing the material leaving the column across the curved surface of the region carries no downward momentum, and the material entering the column across the curved surface also carries no downward momentum. The difference between the momentum flowing in and the momentum flowing out across the plane faces of the region is $\delta h \,.\, d(Qw)/dh$ per unit time. This difference arises from the action of the vertical pressure gradient and of gravity on the material in the region in question. Now the rate at which gravity and the pressure gradient supply downward momentum is $-\delta h(g\rho S + SdP/dh)$ so that

$$(14{\cdot}7)\qquad \frac{d}{dh}(Qw) = -g\rho S - S\frac{dP}{dh}.$$

The equation

$$(14{\cdot}8)\qquad \frac{d}{dh}(Q\mathbf{w}) = -g\boldsymbol{\rho}\mathbf{S} - \mathbf{S}\frac{dP}{dh}$$

can be obtained in a similar way for the upward moving column. Equations (14·7) and (14·8) are the required momentum equations. It may be noted that these equations also form a pair that remains unaltered if the signs of g, h are reversed and symbols in heavy type are interchanged with corresponding symbols in ordinary type.

(*c*) *The adiabatic equations*

The adiabatic equations

$$(14{\cdot}9)\quad Q\left(\frac{dT}{dh}-\frac{2}{5}\frac{T}{P}\frac{dP}{dh}\right)=\boldsymbol{U}(T-\boldsymbol{T})+\frac{w^2}{10\mathfrak{R}}(\boldsymbol{U}-U),$$

$$(14{\cdot}10)\quad Q\left(\frac{d\boldsymbol{T}}{dh}-\frac{2}{5}\frac{\boldsymbol{T}}{P}\frac{dP}{dh}\right)=U(T-\boldsymbol{T})+\frac{\boldsymbol{w}^2}{10\mathfrak{R}}(\boldsymbol{U}-U),$$

can either be deduced *ab initio* or as consequences of (14·4), (14·5), (14·7), and (14·8).

15. The Linear Solution of the Equations of Motion

The independent equations (14·1), (14·2), (14·3), (14·4), (14·5), (14·7), and (14·8) lead, after elimination of S and $\boldsymbol{S}$, to five equations for the eight unknowns w, $\boldsymbol{w}$; T, $\boldsymbol{T}$; U, $\boldsymbol{U}$; P, Q. Thus the equations obtained in section 14 from the laws of conservation of mass, momentum, and energy are not sufficient in themselves to determine a unique solution to the problem. It is not yet certain how far the detailed study of the transfer between the columns would succeed in removing this uncertainty. In the writer's opinion even the most complete set of equations would still permit a wide range of solutions. The reason for this will become clear in the following section.

We now show that the equations of motion have a linear solution of the form

$$(15{\cdot}1)\qquad \left.\begin{aligned} w^2 &= \alpha gh & \boldsymbol{w}^2 &= \boldsymbol{\alpha} gh,\\ \mathfrak{R}T &= \beta gh & \mathfrak{R}\boldsymbol{T} &= \boldsymbol{\beta} gh,\end{aligned}\right\}$$

where α, $\boldsymbol{\alpha}$, β, $\boldsymbol{\beta}$ are constants independent of height. The equations (15·1) together with (14·6) also show that (Qh) is independent of height. It follows that the right-hand sides of equations (14·4) and (14·5) must be zero, so that

$$(15{\cdot}2)\qquad \frac{\boldsymbol{U}}{U}=\frac{\beta+0{\cdot}2}{\boldsymbol{\beta}+0{\cdot}2}.$$

Moreover from (14·2) we see that $(U-\boldsymbol{U})<0$, since $dQ/dh<0$. According to (15·2) this will be the case if $\boldsymbol{\beta}<\beta$. Thus the present solution agrees with the physical condition that the ascending currents must be cooler than the descending currents. The equations (14·2), (15·2), taken together with (Qh) constant, also show that both U, $\boldsymbol{U}$ vary like h^{-2}.

Equations (14·7), (14·8), and (15·1) lead to

$$(15\cdot 3) \qquad -\frac{1}{P}\frac{dP}{dh} = \frac{1 - 0\cdot 5\alpha}{2\beta} = \frac{1 - 0\cdot 5\boldsymbol{\alpha}}{2\boldsymbol{\beta}},$$

which gives a relation between the four constants α, $\boldsymbol{\alpha}$, β, $\boldsymbol{\beta}$ and also shows that we must have $2 > \boldsymbol{\alpha} > \alpha$ in order that both $\boldsymbol{\beta} < \beta$ and $dP/dh < 0$ be satisfied. Thus the ascending columns move faster than the hotter descending columns. We notice further than since the motion of the ascending columns arises from the vertical pressure gradient the velocity $\boldsymbol{w}$ must be less than the velocity of sound. That is, we must have

$$(15\cdot 4) \qquad \boldsymbol{\alpha} \leqslant 10\boldsymbol{\beta}/3.$$

This condition together with $\alpha < \boldsymbol{\alpha}$ also ensures that the descending columns move more slowly than the velocity of sound.

We now specialize the discussion by taking $Ph^{\frac{1}{2}}$ to be a constant, independent of height. The reason for this step becomes clear in the following section. Then (15·3) together with $\boldsymbol{\beta} < \beta$ and $2 > \boldsymbol{\alpha} > \alpha$ gives

$$(15\cdot 5) \qquad 0 < 1 - 0\cdot 5\alpha = \beta > \boldsymbol{\beta} = 1 - 0\cdot 5\boldsymbol{\alpha} > 0.$$

In order that (15·5) be compatible with (15·4) it is necessary that

$$(15\cdot 6) \qquad \boldsymbol{\beta} \geqslant 3/8, \; \boldsymbol{\alpha} \leqslant 5/4.$$

We also have

$$(15\cdot 7) \qquad \beta\rho(gh)^{3/2} = \boldsymbol{\beta\rho}(gh)^{3/2} = 0\cdot 5P(gh)^{\frac{1}{2}} = \text{constant} = K, \text{ say}.$$

The equations (14·1), (15·1), (15·7), taken together with (Qh) constant, give

$$(15\cdot 8) \qquad \left.\begin{aligned} S = \text{constant}, \; \boldsymbol{S} = \text{constant}. \\ S/\boldsymbol{S} = \beta\boldsymbol{\alpha}^{\frac{1}{2}}/\boldsymbol{\beta}\alpha^{\frac{1}{2}} > 1. \end{aligned}\right\}$$

The result (15·8) applies only when $Ph^{\frac{1}{2}}$ is constant. It is this special property that gives exceptional importance to this particular solution. It enables the whole volume of region 1 to be occupied by convection currents, and this is necessary to secure the maximum rate of energy transport. From now on it will be supposed that this situation occurs, so that every element of material present in region 1 takes part in the convective circulation. This question is referred to again in the following section.

The results of the present section are summarized in the following set of equations:

$$(15\cdot9)\qquad \left.\begin{array}{ll} w^2 = \alpha gh, & \boldsymbol{w}^2 = \boldsymbol{\alpha} gh, \\ \mathfrak{R}T = (1 - 0\cdot5\alpha)gh, & \mathfrak{R}\boldsymbol{T} = (1 - 0\cdot5\boldsymbol{\alpha})gh, \\ \rho = K(1 - 0\cdot5\alpha)^{-1}(gh)^{-3/2}, & \boldsymbol{\rho} = K(1 - 0\cdot5\boldsymbol{\alpha})^{-1}(gh)^{-3/2}, \\ P = 2K(gh)^{-\frac{1}{2}}. & \end{array}\right\}$$

The descending currents occupy a fraction

$$(15\cdot10)\qquad \frac{S}{S + \boldsymbol{S}} = \frac{\beta\alpha^{-\frac{1}{2}}}{\beta\alpha^{-\frac{1}{2}} + \boldsymbol{\beta}\boldsymbol{\alpha}^{-\frac{1}{2}}} = \frac{2\boldsymbol{\alpha}^{\frac{1}{2}}(1 - 0\cdot5\alpha)}{(\alpha^{\frac{1}{2}} + \boldsymbol{\alpha}^{\frac{1}{2}})(2 - \alpha^{\frac{1}{2}}\boldsymbol{\alpha}^{\frac{1}{2}})}$$

of the total area, while the ascending currents occupy a fraction

$$(15\cdot11)\qquad \frac{\boldsymbol{S}}{S + \boldsymbol{S}} = \frac{2\alpha^{\frac{1}{2}}(1 - 0\cdot5\boldsymbol{\alpha})}{(\alpha^{\frac{1}{2}} + \boldsymbol{\alpha}^{\frac{1}{2}})(2 - \alpha^{\frac{1}{2}}\boldsymbol{\alpha}^{\frac{1}{2}})}.$$

Equations (15·9), (15·10), (15·11) involve the constants α, $\boldsymbol{\alpha}$, which must satisfy $5/4 > \boldsymbol{\alpha} > \alpha > 0$, and the constant K, which is of dimensions mass/(time)^3.

16. The Fit between Regions 1 and 2

(i) $q \ll 1$.

The formulation of the boundary conditions between regions 1 and 2 is comparatively simple when (15·9), (15·10), and (15·11) remain valid up to the top of region 1. For this to be the case, the total density at the base of region 2 must be large compared with the density of the incoming accreted hydrogen atoms. The important assumption (*c*) of section 13 is then satisfied throughout region 1. It is shown at the end of the present subsection that this requirement is fulfilled when $q \ll 1$.

The rate at which energy is supplied by accretion is

$$0\cdot5 m_p V g q N(r^*) \text{ per cm.}^2$$

The first boundary condition is that the rate of downward transport of energy per cm.^{2} in region 1, as given by dividing the left-hand side of (14·6) by $(S + \boldsymbol{S})$, is equal to the rate of supply of energy by accretion. The expression (14·6), when divided by $(S + \mathbf{S})$ can be written as

$$\frac{1}{(S + \mathbf{S})}\{S\rho w(0{\cdot}5w^2 + 5\mathfrak{R}T) - \mathbf{S}\boldsymbol{\rho}\mathbf{w}(0{\cdot}5\mathbf{w}^2 + 5\mathfrak{R}\mathbf{T})\}$$
$$= \frac{4K\alpha^{\frac{1}{2}}\boldsymbol{\alpha}^{\frac{1}{2}}(\boldsymbol{\alpha}^{\frac{1}{2}} - \alpha^{\frac{1}{2}})}{2 - \alpha^{\frac{1}{2}}\boldsymbol{\alpha}^{\frac{1}{2}}}.$$

Thus

$$(16{\cdot}1) \qquad \frac{4K\alpha^{\frac{1}{2}}\boldsymbol{\alpha}^{\frac{1}{2}}(\boldsymbol{\alpha}^{\frac{1}{2}} - \alpha^{\frac{1}{2}})}{2 - \alpha^{\frac{1}{2}}\boldsymbol{\alpha}^{\frac{1}{2}}} = 0{\cdot}5\ m_p VgqN(r^*)$$

is the first boundary condition.

We next consider the momentum balance for region 2 taken as a whole. The downward momentum flowing into this region due to accretion is $m_p gqN(r^*)$ per cm.^{2} per sec. The average rate at which *downward* momentum enters region 2 at its base is

$$-\mathbf{S}\boldsymbol{\rho}\mathbf{w}^2/(S + \mathbf{S}) \text{ per cm.}^2 \text{ per sec.},$$

while the average rate at which downward momentum leaves region 2 at its base is

$$S\rho w^2/(S + \mathbf{S}) \text{ per cm.}^2 \text{ per sec.}$$

Now from the conservation of momentum the difference between the downward momentum leaving and entering region 2 must be equal to the difference between the action of gravity on the material in this region and the pressure acting at the base of the region. This gives

$$(16{\cdot}2) \quad (S\rho w^2 + \mathbf{S}\boldsymbol{\rho}\mathbf{w}^2)/(S + \mathbf{S}) = m_P N(r^*)g(1 + q) - P.$$

From (15·9), (15·10), (15·11), and (16·2) we obtain

$$(16{\cdot}3) \quad m_P N(r^*)g(1 + q) = \frac{4K\alpha^{\frac{1}{2}}\boldsymbol{\alpha}^{\frac{1}{2}}}{(gh^*)^{\frac{1}{2}}(2 - \alpha^{\frac{1}{2}}\boldsymbol{\alpha}^{\frac{1}{2}})} = \frac{4K(1 - 0{\cdot}5\alpha)^{\frac{1}{2}}}{(\mathfrak{R}T^*)^{\frac{1}{2}}(2 - \alpha^{\frac{1}{2}}\boldsymbol{\alpha}^{\frac{1}{2}})},$$

where h^* is the height of the base of region 2 and T^* is the temperature in a *descending* column at this height. The constant K can be eliminated from (16·1) and (16·3) to give

$$(16{\cdot}4) \quad (\mathfrak{R}T^*)^{\frac{1}{2}} = \frac{Vq}{(1 + q)}\,\frac{(1 - 0{\cdot}5\alpha)^{\frac{1}{2}}}{2\alpha^{\frac{1}{2}}\boldsymbol{\alpha}^{\frac{1}{2}}(\boldsymbol{\alpha}^{\frac{1}{2}} - \alpha^{\frac{1}{2}})} = \frac{Vq}{1 + q}\cdot\zeta, \text{ say.}$$

The value of T^* given by (16·4) provides a more accurate estimate of the temperature in region 2 than the approximate quantity Θ used in section 12. The factor $Vq/(1 + q)$ depends only on the accretion rate, while the factor ζ depends only on the properties of the convection currents in region 1. Now as seen in section 15 there are many arrangements of the currents in region 1 that satisfy the equations of motion. These different arrangements correspond to different values of α, $\boldsymbol{\alpha}$, and K, and consequently to different values of ζ and T^*. We now introduce a criterion for obtaining a unique solution from these different cases. *The solution will be chosen that leads to the smallest value of* T^*. It is plausible to suppose that this condition gives the most stable arrangement of the convection currents since this case gives a minimum total energy for region 2.

The criterion chosen in the previous paragraph requires ζ to be a minimum with respect to both α and $\boldsymbol{\alpha}$. The minimum with respect to $\boldsymbol{\alpha}$ is easily obtained since ζ decreases with increasing $\boldsymbol{\alpha}$. Thus we must choose the highest permissible value of $\boldsymbol{\alpha}$ which by (15·6) is 5/4. When $\boldsymbol{\alpha} = 5/4$ is inserted in ζ it is easily found that the minimum with respect to α occurs at $\alpha = 0{\cdot}38$ and that the minimum of ζ, $\zeta_{\min}$ say, is 1·30. The corresponding minimum of T^*, $T^*_{\min}$ say, is

$$(16{\cdot}5)\qquad \left.\begin{aligned} \mathfrak{R}T^*_{\min} &= 1{\cdot}69\left(\frac{Vq}{1+q}\right)^2, \\ \boldsymbol{\alpha} = 5/4,\ \boldsymbol{\beta} &= 1 - 0{\cdot}5\boldsymbol{\alpha} = 3/8, \\ \alpha = 0{\cdot}38,\ \beta &= 1 - 0{\cdot}5\alpha = 0{\cdot}81. \end{aligned}\right\}$$

The values of β, $\boldsymbol{\beta}$ in (16·5) follow immediately from (15·5). It may be noted that the minimum of ζ with respect to α is very flat. For example, the case $\alpha = 0{\cdot}6$, $\boldsymbol{\alpha} = 5/4$ gives $\zeta = 1{\cdot}40$ which leads to only a seven per cent. increase in T^*. The dependence on $\boldsymbol{\alpha}$, on the other hand, is fairly sensitive. For example, when $\boldsymbol{\alpha} = 1{\cdot}00$ the minimum of ζ with respect to α occurs at $\alpha = 0{\cdot}29$, and is 1·86.

The velocity V is given by the expression $(2GM/r^*)^{\frac{1}{2}}$. We anticipate the value $r^* \sim 8{\cdot}2 \times 10^{10}$ cm. obtained below. Then $V \sim 5{\cdot}7 \times 10^7$ cm./sec. and

$$(16{\cdot}6)\qquad T^*_{\min} = 6{\cdot}6 \times 10^7 \left(\frac{q}{1+q}\right)^2 \ {}^\circ\mathrm{K}.$$

The result (16·6) gives a substantially higher temperature in region 2 than (12·5). The difference arises from including the energy carried by the ascending columns, and moreover the velocity of the descending columns is appreciably less than the velocity of sound. The ratio of the rates of transport of energy in the descending and ascending columns is evidently given by

$$(16\cdot7) \qquad \frac{0\cdot5w^2 + 5\mathfrak{R}T + gh}{0\cdot5\boldsymbol{w}^2 + 5\mathfrak{R}\boldsymbol{T} + gh} = \frac{0\cdot5\alpha + 5\beta + 1}{0\cdot5\boldsymbol{\alpha} + 5\boldsymbol{\beta} + 1} \sim 1\cdot5,$$

when the values given in (16·5) for α, $\boldsymbol{\alpha}$, β, $\boldsymbol{\beta}$ are used. The energy carried by the upward columns is by no means negligible. The values of α, β given in (16·5) show that the velocity of the descending columns is about 0·37 times the velocity of sound. The velocity of the upward currents, on the other hand, is equal to the velocity of sound.

It can be shown that of all relations of the form Ph^n constant, the case $n = 0\cdot5$, which has been used throughout the above discussion, leads to the smallest value of T^*. This justifies the use of this particular case.

The value of the constant K is easily found from (16·1). Putting $\alpha = 0\cdot38$, $\boldsymbol{\alpha} = 5/4$, $V = 570$ km./sec., $g = 2\cdot1 \times 10^4$ cm. sec.$^{-1}$, in this equation gives

$$(16\cdot8) \qquad K = 1\cdot3 \times 10^6 q \text{ gr. sec.}^{-3}.$$

This value of g makes allowance for the base of region 2 being above the photosphere, but the above work is nevertheless slightly approximate on account of the neglect of the dependence of g on h.

The main results for region 1 are

$$(16\cdot9) \qquad \left.\begin{array}{ll} \mathfrak{R}T = 0\cdot81gh, & \mathfrak{R}\boldsymbol{T} = 0\cdot375gh, \\ w^2 = 0\cdot38gh, & \boldsymbol{w}^2 = 1\cdot25gh, \\ \rho = \dfrac{1\cdot6 \times 10^6 q}{(gh)^{3/2}}, & \boldsymbol{\rho} = \dfrac{3\cdot4 \times 10^6 q}{(gh)^{3/2}}, \\ P = \dfrac{2\cdot6 \times 10^6 q}{(gh)^{1/2}}, & \boldsymbol{S} = 4S, \\ \multicolumn{2}{l}{\mathfrak{R}T_{\min} = 0\cdot81gh^* = 6\cdot6 \times 10^7\, \mathfrak{R}q^2/(1+q)^2.} \end{array}\right\}$$

The dependence of $T^*_{\min}$ on q is shown in Table 14.

Table 14.

q	·05	·10	·15	·20	·25	·50
$T^*_{\min}$ (10^6 °K. as unit)	·15	·55	1·1	1·8	2·7	7·4

The quantity q is the only free parameter in the present theory. For comparison with observation it is necessary to assign some value to q. *It is shown in chapter* IV *that* $q = \cdot 25$ *leads to results that are in good agreement with observation.* With this value of q, (16·9) can be written in the form

$$\left.\begin{aligned}
&T = 2{\cdot}7 \times 10^6 h/h^* \text{ deg. K.}, \qquad \mathcal{T} = 1{\cdot}2 \times 10^6 h/h^* \text{ deg. K.},\\
&w = 101(h/h^*)^{\frac{1}{2}} \text{ km./sec.}, \qquad \mathfrak{w} = 184(h/h^*)^{\frac{1}{2}} \text{ km./sec.},\\
&\rho = 8{\cdot}9 \times 10^{-17}(h^*/h)^{3/2} \text{ gr./cm.}^3\\
&\qquad = 5{\cdot}3 \times 10^7(h^*/h)^{3/2} \text{ atoms/cm.}^3,\\
&\boldsymbol{\rho} = 1{\cdot}9 \times 10^{-16}(h^*/h)^{3/2} \text{ gr./cm.}^3\\
&\qquad = 1{\cdot}2 \times 10^8(h^*/h)^{3/2} \text{ atoms/cm.}^3,\\
&P = 3{\cdot}9 \times 10^{-2}(h^*/h)^{\frac{1}{2}} \text{ dynes/cm.}^2,\\
&h^* = 1{\cdot}3 \times 10^{10} \text{ cm.}
\end{aligned}\right\} \tag{16·10}$$

These values can be applied to the considerations of section 12. The velocity of the downward currents in region 2 relative to the incoming accreted material is $\sim$470 km./sec., whereas the relative velocity of the upward currents is $\sim$750 km./sec. Thus the ratio of the collision cross-sections between incoming accreted material and material in the descending and ascending columns respectively $\sim(750/470)^4$ which is $\sim 7/1$. This result, together with $S/\mathcal{S} = 4$, means that the descending material in region 2 takes up almost the whole of the momentum of the incoming material. This conclusion provides strong support for the theory of convective instability discussed in section 12.

It can be shown from (16·9) that the density of the solar atmosphere at height h^* is $\sim 3 \times 10^6 \,(1+q)^3/q^2$ atoms/cm.^{3}, whereas (11·8) gives a density $\sim 8{\cdot}3 \times 10^6 q$ atoms/cm.^{3} for the incoming accreted material. Thus when $q \ll 1$ (for example, $q = \cdot 25$) the density of the solar atmosphere at the top of region 1 is large compared with the density of the incoming atoms. This result confirms the remarks of the first paragraph of the present subsection.

(ii) $q > 1$.

When $q > 1$ the density of the incoming accreted atoms is comparable with the density of the solar atmosphere at the top of region 1. In this case, assumption (c) of section 13 is not a good approximation near the top of region 1. Consequently, the method given in (i) leads to an inaccuracy in the determination of T^*_{min}. It can be shown that a better approximation for T^*_{min} is obtained by equating $3\mathfrak{R}T^*_{\mathrm{min}}$ with the energy per unit mass $0{\cdot}5V^2$ of the incoming accreted material. The following solution may then be employed as giving order of magnitude estimates for T, $\boldsymbol{T}$, w, $\mathfrak{w}$, ρ, $\boldsymbol{\rho}$, P, h^*, and T^*_{min}:

$$(16{\cdot}11)\qquad \left.\begin{aligned} &\mathfrak{R}T = 0{\cdot}81gh, && \mathfrak{R}\boldsymbol{T} = 0{\cdot}375gh, \\ &w^2 = 0{\cdot}38gh, && \mathfrak{w}^2 = 1{\cdot}25gh, \\ &\rho = \frac{1{\cdot}6\times 10^6 q}{(gh)^{3/2}}, && \boldsymbol{\rho} = \frac{3{\cdot}4\times 10^6 q}{(gh)^{3/2}}, \\ &P = \frac{2{\cdot}6\times 10^6 q}{(gh)^{\frac{1}{2}}}, && \boldsymbol{S} = 4S, \\ &\mathfrak{R}T^*_{\mathrm{min}} = 0{\cdot}81gh^* = V^2/6. && \end{aligned}\right\}$$

CHAPTER IV

THE CHROMOSPHERE AND CORONA: THEORY PART II

17. Radiation in Region 1

(i) *Emission by hydrogen*

The most important contribution to the radiation emitted by the material in region 1 arises from hydrogen. The radiating processes are: (*a*) electron-proton recombinations to the ground state of the hydrogen atom, (*b*) electron-proton recombinations to the excited states of the hydrogen atom, (*c*) free-free electronic transitions, and (*d*) the excitation of neutral atoms by electron collisions.

Near the base of region 1, which is the main emitting zone, (*b*), (*c*), and (*d*) may be neglected in comparison with (*a*). Then the rate of radiation in descending material is given, in the frequency range ν to $(\nu + d\nu)$, by multiplying (7·7) by n_p and putting $Z = 0$, $s = 1$. Thus we obtain

(17·1) $$2 \cdot 12 \times 10^{-32} e^{(x-h\nu)/kT} n_e n_p \,.\, d\nu / T^{3/2}, \quad (h\nu > x),$$

where $x = 2 \cdot 15 \times 10^{-11}$ ergs is the ionization potential of the ground state of hydrogen.

Consider a unit (cross-section 1 cm.^{2}) radial column lying within a *descending* current, and let the material at the base of the column possess temperature T_0. For the moment, we suppose that the hydrogen is effectively wholly ionized. Then, taking the assembly to be electrically neutral, we obtain from (16·9) the equations

(17·2) $$n_e = n_p = \frac{1 \cdot 6 \times 10^6 q}{m_p (gh)^{3/2}} = \frac{9 \cdot 4 \times 10^{17} q}{T^{3/2}},$$

(17·3) $$\frac{dT}{dh} = \frac{0 \cdot 81 g}{\mathfrak{R}} = 2 \cdot 6 \times 10^{-4}.$$

It can be shown from (17·1), (17·2), (17·3) that with a slight approximation the rate of radiation, in the frequency range ν to $(\nu + d\nu)$, by the material in the unit column is

$$(17\cdot 4) \qquad 7\cdot 0 \times 10^7 q^2 \,.\, d\nu \int_{T_0}^{\infty} \frac{dT}{T^{9/2}} e^{(x-h\nu)/kT} \text{ ergs/sec., } (h\nu > x).$$

The approximation arises from taking ∞ as the upper limit of the integral.

If we take a similar unit column in *ascending* material, with base at the same level as the base of the column considered in the previous paragraph, then the temperature $\boldsymbol{T}_0$ corresponding to T_0 is given by

$$(17\cdot 5) \qquad \boldsymbol{T}_0 = 0\cdot 463\, T_0.$$

The corresponding emission by the ascending material can be shown to be

$$(17\cdot 6) \qquad 7\cdot 0 \times 10^7 q^2 \,.\, d\nu \int_{\boldsymbol{T}_0}^{\infty} \frac{d\boldsymbol{T}}{\boldsymbol{T}^{9/2}} e^{(x-h\nu)/k\boldsymbol{T}} \text{ ergs/sec.}$$

The total emission is obtained from (17·4) and (17·6) for ascending and descending material respectively, by integrating with respect to ν from x/h to ∞. Remembering that $S/\boldsymbol{S} = 4$, the *average* emission for ascending and descending material taken together is given by

$$(17\cdot 7)$$

$$7\cdot 0 \times 10^7 q^2 \left\{ \tfrac{4}{5} \int_{T_0}^{\infty} \frac{dT}{T^{9/2}} \int_{x/h}^{\infty} d\nu\, e^{(x-h\nu)/kT} + \tfrac{1}{5} \int_{\boldsymbol{T}_0}^{\infty} \frac{d\boldsymbol{T}}{\boldsymbol{T}^{9/2}} \int_{x/h}^{\infty} d\nu\, e^{(x-h\nu)/k\boldsymbol{T}} \right\},$$

which can be reduced to

$$1\cdot 9 \times 10^{17} q^2 / \boldsymbol{T}_0^{5/2} \text{ ergs/cm.}^2\text{/sec.,}$$

by using (17·5). Now in order that a photon may escape from the solar atmosphere it is necessary (*a*) that the direction of emission be outwards, (*b*) that there be no absorption in overlying material. For the moment we omit the question raised by (*b*), but we reduce (17·7) by a factor 0·5 to allow for (*a*). Thus the rate of radiation in outward directions, by material lying above a height corresponding to $\boldsymbol{T}_0$, is of order

$$(17\cdot 8) \qquad 9\cdot 4 \times 10^{16} q^2 / \boldsymbol{T}_0^{5/2} \text{ ergs/cm.}^2\text{/sec.}$$

The work of chapter III is based on the assumption that the rate of loss of energy, through radiation occurring in region 1, is small compared with the rate of supply of energy by accretion. This requires (17·8) to be less than $0{\cdot}5 m_p V g N(r^*)$, which is equal to $1{\cdot}2 \times 10^6 q$ ergs/cm.²/sec., when $g = 1{\cdot}9 \times 10^4$ cm./sec.$^{-2}$, $V = 570$ km./sec., and $N(r^*) = 1{\cdot}36 \times 10^{18}$ atoms/cm.² This condition is satisfied if

$$T_0 > 2{\cdot}3 \times 10^4 q^{2/5} \text{ deg. K.,} \tag{17·9}$$

which leads to the values shown in Table 15. This result is very satisfactory, for it is seen that in the case $q = {\cdot}25$ the assumption

Table 15.

q	·1	·25	·5	·75	1	2	5	10	100	1000
T_0 (10^4 deg. K. as unit) >	·90	1·3	1·7	2·0	2·3	3·0	4·3	5·7	14	36

of chapter III is satisfied down to layers immediately overlying the reversing layer (see section 18). Thus for $q = {\cdot}25$ the assumption of negligible radiation holds down to values of T_0 near $1{\cdot}3 \times 10^4$ deg. K., which is comparable with the temperature of the reversing layer. Moreover, this conclusion applies *a fortiori*, since neutral atoms give appreciable absorption, so that only a fraction of the outward moving quanta succeed in escaping from the solar atmosphere. This again raises the condition (*b*), mentioned above, which will now be discussed.

Radiation is not the most important question affecting the validity of (16·9) near the base of region 1. Thus, when the concentration of neutral atoms is appreciable, account must be taken of the difference of ionization energy between ascending and descending material. This effect is similar to the process discussed by Eddington in connexion with convective instability below the photosphere (see section 5), and must operate so as to enable neutral hydrogen atoms to rise to considerably greater heights than is obtained by applying ionization equilibrium to the values of T, T given in (16·9). Thus it is to be expected that neutral hydrogen atoms will occur in appreciable concentration up to heights where

$$1{\cdot}5 kT \sim x.$$

With this condition the radiation emitted in accordance with (17·8) cannot be taken as escaping from the sun unless $T_0 > 4 \times 10^4$ deg. K. (ionization equilibrium applied to (16·9) gives

$$T_0 > 2 \times 10^4 \text{ deg. K.).}$$

Accordingly the average rate at which quanta with energies $>13{\cdot}5$ eV. escape from the sun (due to emission by hydrogen), is given approximately by putting $T_0 = 4 \times 10^4$ deg. K. in (17·8). For $q = {\cdot}25$ this gives an average emission $\sim 10^4$ ergs/cm.²/sec. Moreover, the quanta have a distribution, with respect to frequency, of the form given by (17·6). That is, the emitted spectrum depends on ν according to the factor

$$d\nu \,.\int_{T_0}^{\infty} \frac{dT}{T^{9/2}}\, e^{(\chi - h\nu)/kT},$$

which can be reduced to

$$(17{\cdot}10) \qquad \frac{d\nu}{T_0^{7/2}}\left\{\frac{15}{4x_0^7}\int_0^{x_0} e^{-t^2}\,dt - e^{-x_0^2}\left(\frac{15}{4x_0^6} + \frac{5}{2x_0^4} + \frac{1}{x_0^2}\right)\right\},$$

where $x_0^2 = (h\nu - \chi)/kT_0$. The expression in brackets gives the form of the frequency spectrum and is plotted in Fig. 6 for $T_0 = 4 \times 10^4$ deg. K.

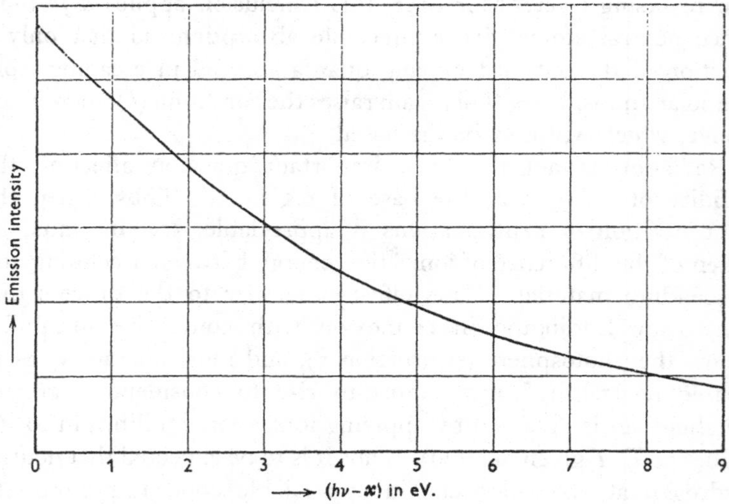

Fig. 6. The area under the curve corresponds to an emission rate $\sim 10^4$ ergs per cm.² per sec. The ionization potential $\chi = 13{\cdot}53$ eV.

(ii) *Emission by heavy elements*

Although the *total* rate of radiation by hydrogen is large compared with the emission by heavy positive ions, the contribution of hydrogen to the high energy part of the spectrum is small compared with that of the heavy elements (*Fe* for example). Thus the main hydrogen emission occurs near the base of region 1, where the emitted quanta have energies of the order of, but greater than, 13·53 eV. The heavy positive ions, on the other hand, give an appreciable contribution at temperatures $\sim 5 \times 10^5$ deg. K. on account of the importance of the $(Z + 1)^4$ factor in (7·7). The values in Table 11 show that the ionization potentials are ~ 300 eV. when $Z \sim 10$. Taking *Fe* as an example, we now examine the spectrum of the high energy quanta emitted in the recombination of electrons with atoms possessing ionization potentials of this order.

The energy of a quantum emitted in the recombination of an electron and a positive ion is the sum of (*a*) the ionization potential of the atom formed by the recombination, (*b*) the kinetic energy of the captured electron. The contribution of (*b*), in the case of heavy ions at high temperature, is smaller than (*a*) (the opposite situation occurs for hydrogen at high temperatures). For example, the mean thermal energy of an electron at a temperature of $1{\cdot}2 \times 10^6$ deg. K. is ~ 150 eV., which is only about one third of the value of x corresponding to this temperature, as given by Table 11. It follows that quanta emitted at high temperature in the recombination of electrons and positive ions have energies of the same order as the ionization potential of the atom formed by the recombination. Thus if (7·7) is integrated with respect to ν from x/h to ∞ we obtain

$$4{\cdot}43 \times 10^{-22}(Z + 1)^4 n_e / s^3 T^{3/2} \tag{17·11}$$

for the energy emitted at temperature T per ion of charge $(Z + 1)$, *and this emission is composed of quanta with energies* $\sim x$, where x is the ionization potential of an atom with Z electrons stripped. The charge Z is interpreted throughout the following discussion as given in terms of T by Table 11. Then if $c_{Fe} n_p$ is the number of *Fe* atoms per cm.^{3} in *descending* material, the total emission per cm.^{3} is

$$4{\cdot}43 \times 10^{-22} c_{Fe}(Z + 1)^4 n_e n_p / s^3 T^{3/2}. \tag{17·12}$$

The range of the emitted spectrum, of special interest in relation to chapter VI, is from ~ 300 eV. to ~ 400 eV. In this range the values given in Table 11 can be represented by the relations

(17·13) $$T = 1{\cdot}9 \times 10^{21} x^{5/3},$$

(17·14) $$Z + 1 = 2{\cdot}3 \times 10^{10} x,$$

where x is expressed in ergs. By using (17·3), (17·13), (17·14), together with the expression (17·12), it can be shown that the material in a unit radial column gives an emission rate for quanta with energies lying between x and $(x + dx)$ of order

(17·15) $$1{\cdot}6 \times 10^5 c_{Fe} q^2 dx/x^{7/6} \text{ ergs per cm.}^2 \text{ per sec.},$$

where s is put equal to 3 over the range of x in question. It can be shown similarly that the emission spectrum from a unit column in *ascending* material is given by

(17·16) $$1{\cdot}6 \times 10^5 \mathbf{c}_{Fe} q^2 dx/x^{7/6} \text{ ergs per cm.}^2 \text{ per sec.},$$

where $\mathbf{c}_{Fe} n_p$ is the number of *Fe* atoms/cm.^{3} in the ascending material (we include the possibility discussed in chapter VI that c_{Fe} may differ from $\mathbf{c}_{Fe}$).

It is emphasized that (17·14) is not valid for $x > 487$ eV. This is due to the change of the principle quantum number s from 3 to 2 as Z increases from 15 to 16. As shown by (7·8) and Table 11 the corresponding increase of x is so large that Z can only increase very slowly with the temperature when this stage is reached. Accordingly, the factor $(Z + 1)^4$ in (7·12) must then be taken as approximately constant. Hence for $x > \sim 500$ eV. the spectrum is approximately of the form dx/x^5.

At the beginning of subsection (i) we noted that the main contribution to the radiation by hydrogen is due to electron-proton recombinations. In principle, this result arises from the fact that the potentials of the excited states of hydrogen are comparable with the ionization potential. A different situation occurs for heavy atoms with many electrons stripped, for, as shown in Table 7, the ionization potentials are very large compared with the excitation potentials of the visible lines in the coronal emission spectrum. In contrast with the case of hydrogen the observed intensity in these lines is comparable with the emission from recombinations of free electrons and the heavy ions. Unfortunately, the cross-section for the excitation of these lines by electron impact is difficult to estimate, because in the case of interest the mean kinetic energy of the electrons is large compared with the excitation potentials.† When an accurate value of this cross-section becomes available it will be possible to estimate c_{Fe}, c_{Fe} by comparing the calculated and observed intensities of the coronal emission lines. In this connexion we note that, in spite of the low excitation potentials of many of the lines, there is little contribution to the coronal emission spectrum due to excitation by quanta emitted from the photosphere. This arises from the forbidden nature of the transitions (see the probabilities in Table 7).

18. The Fit of Region 1 on to the Reversing Layer

The reversing layer lies between the photosphere and the base of region 1. This region, which would exist even if there were no accretion of interstellar material, has a roughly uniform temperature of $\sim 4830°$ K. Assuming the hydrogen to be neutral, the scale height ~ 146 km. (see section 6), and the depth of the layer is

$$\sim 146 \ln(n'/n'') \text{ km.}, \tag{18·1}$$

where n', n'' are the densities of hydrogen atoms at the photosphere and at the base of region 1 respectively. The density n' is $\sim 10^{16}$ atoms/cm.^{3}

Although (16·10) does not represent an accurate solution at temperatures $< \sim 2 \times 10^4$ deg. K. we may nevertheless use these results as giving a general order of magnitude. Then the continuity of density and temperature at the interface between the reversing layer and region 1 leads to an estimate for n''. Thus (16·10) shows that when $q = \cdot 25$ the hydrogen density $\sim 5 \times 10^{11}$ atoms/cm.^{3} when the temperature in region 1 falls to a value comparable with that of the reversing layer. Accordingly, we have $n' \sim 10^{16}$ atoms/cm.^{3}, $n'' \sim 5 \times 10^{11}$ atoms/cm.^{3}, and by (18·1) the top of the reversing layer is at a height ~ 1500 km. above the photosphere (the uncertainty in n'' does not appreciably affect this value, since n'' appears only in the logarithmic term). The present result agrees with the identification given in section 6.

Finally, we note that the co-ordinate h, used in chapter III, has origin at ~ 300 km. below the base of region 1.

† The collisions are of a type similar to the 'triplet' excitations discussed in N. F. Mott and H. S. W. Massey, *The Theory of Atomic Collisions*, Oxford, 1933, p. 192.

19. The Theoretical Electron Densities for $q = \cdot 25$

(i) *At the base of region* 1

The density at the base of region 1 is $\sim 5 \times 10^{11}$ atoms/cm.^{3} and the temperature is $\sim 5000°$ K. Although at this temperature ionization of hydrogen due to electron collisions may be neglected, an important effect arises from the emission of radiation at higher levels. According to section 17 the *downward* emission of quanta with energies $> 13\cdot 53$ eV. by material in region 1 at temperatures $\sim 2 \times 10^4$ deg. K. $\sim 10^5$ ergs/cm.^{2}/per sec. This flux produces ionization in the cooler layers at the base of region 1.

To estimate the quantitative effect of this process we note that the flux of ionizing quanta emitted by the photosphere is only $\sim 5 \times 10^2$ ergs/cm.^{2}/sec. A photospheric temperature $\sim 7200°$ K. would be required to give an equivalent flux. Accordingly, it is to be expected that the degree of ionization at the base of region 1 will be given, so far as order of magnitude is concerned, by the thermodynamic value corresponding to $\sim 7200°$ K. Anticipating that $P_e \sim \cdot 25$ dynes/cm.^{2} it is seen from Table 8 that the hydrogen is then about 50% ionized. This gives an electron density at the base of region 1 $\sim 2\cdot 5 \times 10^{11}$ per cm.^{3}.

(ii) *At greater heights in region* 1

At temperatures $> \sim 2 \times 10^4$ deg. K. the hydrogen is largely ionized, and the electron densities are given by (16·10). Remembering the results of section 18, which show that the origin of the co-ordinate h is at a height of ~ 1200 km. above the photosphere, we obtain the approximate values given in Table 16 for the electron densities.

Table 16.

Height above photosphere (km.)	3500	10,000	50,000	130,000
Electron density (per cm.^{3})	5×10^{10}	5×10^9	5×10^8	10^8

(iii) *In region* 2

There are two contributions to the density of hydrogen atoms in region 2:

(*a*) For simplicity we assume that $\mathfrak{R}\Theta/g(1+q)\mu$ is constant throughout region 2, and at the base of region 2 we put the temperature Θ equal to 2×10^6 deg. K., $g = 1{\cdot}9 \times 10^4$ cm. sec.$^{-2}$, $q = {\cdot}25$, $\mu = 0{\cdot}5$. Then the scale height throughout the material belonging to the solar atmosphere is everywhere equal to $1{\cdot}4 \times 10^{10}$ cm. Accordingly, the density of the solar atmosphere in region 2 is given approximately by

$$(19{\cdot}1) \qquad 10^8 \exp\{5(1{\cdot}2 - r)\} \text{ atoms/cm.}^3,\ r > 1{\cdot}2,$$

where r is the distance from the solar centre in terms of $R(6{\cdot}9 \times 10^{10}$ cm.) as unit. The distance $r^* = 1{\cdot}2$ corresponds to the base of region 2. Since the atoms of the solar atmosphere are effectively wholly ionized at a temperature $\sim 2 \times 10^6$ deg. K., (19·1) gives the contribution of the solar atmosphere to the electron density in region 2.

(*b*) There is also a contribution from the incoming accreted hydrogen atoms. Remembering that these atoms are moving radially towards the sun with velocity $\sim 617/r^{\frac{1}{2}}$ km./sec. ($r > 1{\cdot}2$), it follows from the hydrodynamic equation of continuity that their density is proportional to $r^{-3/2}$. Furthermore (11·8) gives a density $\sim 2{\cdot}1 \times 10^6$ atoms per cm.^{3} at distance $r^* = 1{\cdot}2$ in the case $q = 0{\cdot}25$. Thus the density at distance r is

$$(19{\cdot}2) \qquad 3{\cdot}0 \times 10^6/r^{3/2} \text{ atoms/cm.}^3,\ (r > 1{\cdot}2).$$

Table 17 gives the contributions of (19·1) and (19·2) to the hydrogen density at various distances from the sun.

Table 17.

r	1·2	1·3	1·4	1·6	1·8	2·0	2·2	2·4	2·6	2·8	3·0	3·5	4	5	6	8	10
log (19·1)	8·0	7·8	7·6	7·1	6·7	6·3	5·8	5·4	5·0	4·5	4·1						
log (19·2)	6·4	6·3	6·3	6·2	6·1	6·0	6·0	5·9	5·9	5·8	5·8	5·7	5·6	5·4	5·3	5·1	5·0
log {(19·1) + (19·2)}	8·0	7·8	7·6	7·2	6·8	6·5	6·2	6·0	5·9	5·8	5·8	5·7	5·6	5·4	5·3	5·1	5·0

Although the form of (19·1) depends on the assumption of a constant scale height, and the values given in Table 17 are consequently approximations, the general features shown by this table are nevertheless of importance. In particular, the contribution of the solar atmosphere to the hydrogen density falls off with increasing r much more rapidly than the density of the incoming accreted material. For example, according to Table 17, the solar atmosphere makes the main contribution to the total density when $r < 2$, whereas for $r > 2$ the accreted atoms supply the main contribution. Indeed, beyond $r = 2{\cdot}5$ the extension of the solar atmosphere is weak.

The accreted atoms may become ionized even at large distances from the sun owing to the absorption of ionizing quanta. In section 17 it was shown that the rate of emission of outward moving quanta with energies $>13{\cdot}5$ eV. is $\sim 10^4$ ergs/cm.^{2}/sec. This corresponds to an outward flux $\sim 4 \times 10^{14}$ ionising quanta/cm.^{2}/sec. Since this value is appreciably greater than the rate $\sim 10^{14}$/cm^2/sec. at which accreted atoms enter the sun, it follows that the effect of absorption on the ionizing radiation may be neglected in order of magnitude considerations (the recombination of electrons and protons in the outer corona can be neglected). Thus the outward flux of ionizing quanta is of order

$$(19{\cdot}3) \qquad 4 \times 10^{14}/r^2 \text{ per cm.}^2 \text{ per sec.}$$

at distance r.

Now the time required for an incoming atom to move from $(r + dr)$ to r is $1{\cdot}12 \times 10^3 r^{\frac{1}{2}} dr$ sec. The probability of a neutral atom being ionized during this motion is therefore given on multiplying (19·3) by $1{\cdot}12 \times 10^3 \sigma r^{\frac{1}{2}} dr$, where σ is the average cross-section for absorption of the ionizing quanta. Thus the probability of a neutral atom becoming ionized *before* it reaches the distance r is

$$(19{\cdot}4) \quad 1 - \exp\left\{-4{\cdot}5 \times 10^{17}\sigma \int_r^\infty \frac{dy}{y^{3/2}}\right\} = 1 - \exp\left\{-\frac{9 \times 10^{17}\sigma}{r^{\frac{1}{2}}}\right\}.$$

The cross-section for absorption of a quantum of energy $h\nu(>x)$ is $6{\cdot}28 \times 10^{-18}(x/h\nu)^3$, where x (13·53 eV.) is the ionization potential of the ground state of the hydrogen atom. In the present problem the main contribution to the flux (19·3) arises from quanta with energies only slightly exceeding 13·53 eV. Thus we may put $\sigma \sim 5 \times 10^{-18}$ in (19·4). We then obtain the probabilities shown in Table 18. The probabilities are sensitive to the numerical

Table 18.

r	2	5	10	100
Probability of a neutral atom becoming ionized before reaching distance r	·96	·87	·76	·36

factor in (19·3). Thus if the flux of ionizing quanta is taken as

(19·5) $8 \times 10^{14}/r^2$ per cm.^{2} per sec.,

the probabilities are altered to Table 19. Since only the order of magnitude of the flux of ionizing quanta is known, the probabilities

Table 19.

r	2	5	10	100
Probability of a neutral accreted atom becoming ionized before reaching distance r	$>$·99	·98	·94	·60

given by the present theory must evidently be applied with caution. It seems probable, however, that the uncertainties do not affect the conclusion that the incoming hydrogen atoms are largely ionized before reaching $r = 10$. This result is important, for when hydrogen is wholly ionized the electron density may be taken as equal to the density of hydrogen atoms. Consequently the values given in Table 17 can be interpreted as electron densities.

(iv) *Comparison of the theoretical and observed electron densities*

The theoretical results given in (i), (ii), (iii) are compared in Table 20 with the observed electron densities (see sections 6 and 8). The satisfactory agreement between the observed and theoretical values, over the range from 1500 km. above the photosphere

Table 20.

$q = \cdot 25$

Height above photosphere (km.)	600	1500	3500	10,000
$(n_e)_{\text{observed}}$, as given by Balmer continuum	2×10^{11}	$\sim 10^{11}$	7×10^{10}	
$(n_e)_{\text{observed}}$ as given by Stark effect	3×10^{11}			
$(n_e)_{\text{theoretical}}$		$\sim 2{\cdot}5 \times 10^{11}$	5×10^{10}	5×10^{9}

r (solar radii)	1·1	1·2	1·3	1·4	1·6	1·8	2·0	2·2	2·4	2·6	2·8	3·0	3·5	4	5	6	8	10
$(\log n_e)_{\text{observed}}$	8·2	7·8	7·6	7·4	7·0	6·8	6·5	6·4	6·2	6·1	6·0	6·0	5·8	5·7	5·6	5·4	5·2	5·0
$(\log n_e)_{\text{theoretical}}$	8·6	8·0	7·8	7·6	7·2	6·8	6·5	6·2	6·0	5·9	5·8	5·8	5·7	5·6	5·4	5·3	5·1	5·0

out to 10 solar radii, provides strong evidence in favour of the case $q = \cdot 25$.

A plot of the electron density up to a height of 9000 km. is given in Fig. 7. The values used in drawing this curve were obtained from (16·10) for heights > 2500 km., but observational data was also taken into account for heights < 2500 km.

20. The Chromospheric Spectrum

(i) *The metallic lines*

Let n_Z, $Z = 0, 1, 2, \ldots$, denote the densities of atoms of a given metal that are neutral, in the first ionization stage, in the

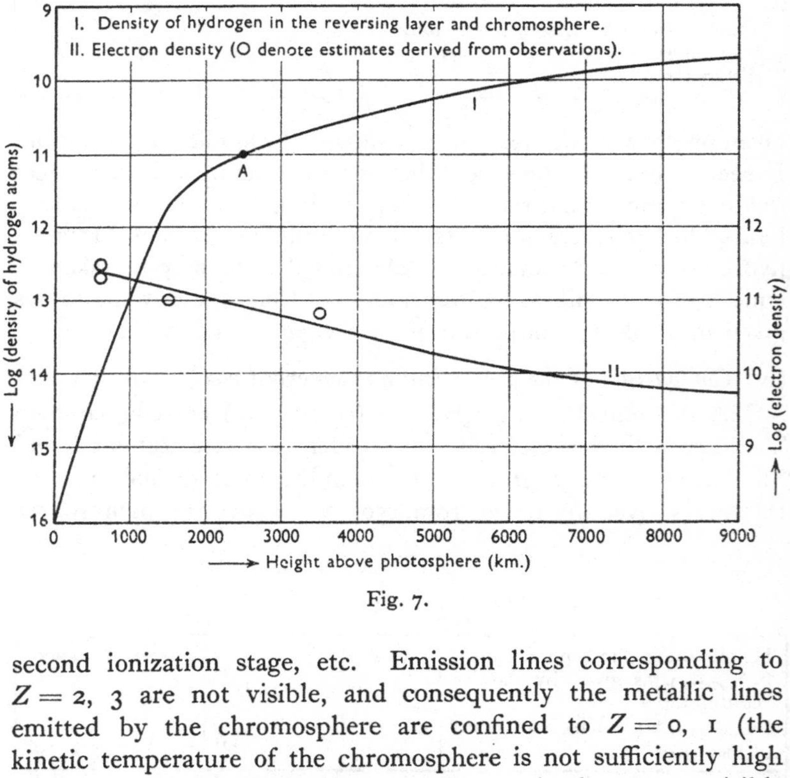

Fig. 7.

second ionization stage, etc. Emission lines corresponding to $Z = 2, 3$ are not visible, and consequently the metallic lines emitted by the chromosphere are confined to $Z = 0, 1$ (the kinetic temperature of the chromosphere is not sufficiently high to give atoms with Z of order ten, when, as in the corona, visible lines are again emitted).

First, we consider the ratio n_0/n_1 as a function of height. Ionization by the photoelectric effect is important for the case $Z = 0$ on account of the low ionization potentials of neutral metals (< 8 eV.). By a straightforward calculation using (*a*) the cross-section (7·5) for ionization by electron collision, (*b*) a cross-section $\sim 2 \times 10^{-17}$ for photoelectric ionization, (*c*) the assumption that the photosphere radiates in the ultraviolet like a black body, it can be shown that the photoelectric effect is the main ionization process at temperatures $< \sim 1{\cdot}5 \times 10^4$ deg. K. That is, for heights $< \sim 2500$ km., we may consider the ionization of neutral metals on the basis of an equation with form similar to (7·3). Now from Fig. 7 it is seen that the electron density varies slowly in this range of height. Accordingly, the ratio n_0/n_1 decreases only slowly with increasing height below 2500 km. A comparatively rapid decrease above this level arises, however, from the increasing importance of electron collisions.

Next, we examine the variation with height of the ratio n_2/n_1. The ionization potentials of singly ionized metals are comparable with the ionization potential of hydrogen (see Table 5). Thus, as in the case of hydrogen, ionization by electron collisions is more effective than radiative ionization. The ionization equilibrium between $Z = 1$ and $Z = 2$ is therefore determined by (7·10), which leads to results essentially similar to those already obtained for hydrogen. Thus, we expect n_1/n_2 to be either large or small, according as the kinetic temperature is appreciably less than, or greater than, $\sim 1{\cdot}5 \times 10^4$ deg. K.

The remarks of the preceding paragraphs show that $n_2, n_3, \ldots,$ are small compared with n_0, n_1, and that n_0/n_1 is approximately independent of height below 2500 km., whereas above 2500 km. both n_0, n_1 rapidly become small compared with n_2. It follows, since the chromospheric lines correspond to $Z = 0, 1$ that there must be a rapid decrease in the intensities of these lines at a height of about 2500 km. It is therefore a satisfactory feature that, out of about three thousand metallic lines observed, there is only the small group shown in Table 21 remaining visible above this height.

The great height attained by the *H*- and *K*-lines of *Ca* II suggest that the lines appearing in Table 21 are exceptional in that the above considerations of ionization equilibrium do not apply in these cases. The association of corpuscular emission by the sun with solar flares (see chapter VI) indicates that metallic atoms possessing strong lines with excitation potentials ~ 4 eV. are powerfully repelled by radiation pressure. It is therefore possible that the atoms listed in Table 21 are being continually sprayed radially outwards with speeds that are large compared with the convective motions in the chromosphere. Under normal conditions it is likely that these atoms eventually become ionized and fall back to lower regions. The observed emission above 2500 km. is attributed to the disturbance of ionization equilibrium arising from this process.

The variation of intensities of metallic lines *below* 2500 km. can be obtained by using (*a*) the result obtained above, that n_2, n_3, . . . , are small compared with n_1, and that n_0/n_1 is approximately

Table 21.

Atom	Line (Angströms)	Height attained (km.)	Excitation potential (eV.)
Ti II	3685·25	6000	3·92
Ti II	3759·33	6000	3·89
Ti II	3761·33	6000	3·85
Mg I	3829·35	5000	5·92
Mg I	3832·34	5000	5·92
Mg I	3838·30	6000	5·92
Ca II	3933·90	14,000	3·14
Ca II	3968·70	14,000	3·11
Sr II	4077·83	6000	3·03
Sr II	4215·70	6000	2·93
Sc II	4226·74	4000	2·92
Sc II	4246·90	5000	3·22

independent of height, (*b*) the assumption that the relative abundances of metals and hydrogen are independent of height, (*c*) the following thermodynamic considerations. The excitation potentials of the metallic lines are only a few electron volts. Consequently, the main solar spectrum, arising from the emission by the photosphere, provides a powerful flux of quanta capable of populating the excited states corresponding to the observed lines. It can be shown that because the chromospheric lines arise from allowed transitions, this process outweighs the effect of a high kinetic temperature (this is in contrast with the case of hydrogen). Thus we expect the populations in the excited states to approximate to the thermodynamic values corresponding to the effective temperature of the radiation at the resonant wavelengths. On account of the darkening in the Fraunhofer lines this effective temperature is appreciably less than the temperature of the photosphere, and is usually taken as $\sim$4000 deg. K.

It follows from (*a*) and (*c*) that below 2500 km. the line intensities have the same dependence on height as the density of the metal in question. Then, by (*b*), it is seen that *the intensities must have the same variation with height as the hydrogen density.* This dependency is shown by curve I in Fig. 7, and the limiting height of 2500 km. is marked by the point *A*.

The suggestion made above, that the exceptional lines in Table 21 involve a breakdown of ionization equilibrium, receives support from Wildt's analysis † of observational data for *Sc* II and *Ti* II. Thus, ignoring the limiting point *A*, the curve I of Fig. 7 has a form very similar to the figures obtained by Wildt.

(ii) *The hydrogen lines*

The present theory depends on the absorption and emission properties of Lyman α-quanta.

(*a*) *Conversion of the thermal energy of material into Lyman α-quanta*

Lyman α-quanta are produced by transitions from the $s = 2$ to the $s = 1$ state of the neutral hydrogen atom. Electron-proton recombinations into the $s = 2$ states give the main contribution to the rate of conversion of thermal energy into Lyman α-quanta. According to (7·9) the rate of this process is about half of the recombination rate into the $s = 1$ state. Thus, the rate of conversion into Lyman α-quanta, of energy possessed by the ionized hydrogen lying in a unit radial column extending outwards from the base of the chromosphere, is of the same order as the rate of emission of quanta with energies $> 13{\cdot}53$ eV. The latter emission was shown in section 17 to be $\sim 10^5$ ergs per cm.^{2} per sec.

This emission is enormously greater than the rate of radiation of Lyman α-quanta from a black body at the temperature of the photosphere.

(*b*) *The conversion of Lyman α-quanta into thermal energy*

A Lyman α-quantum must continue to be absorbed and re-emitted by neutral hydrogen atoms until either it escapes from the solar atmosphere or is lost through the ionization of the absorbing atom. The latter process, which is the inverse of (*a*), consists in the Lyman α-quantum being first absorbed by a neutral atom,

† R. Wildt, *Ap.J.*, **105**, 36, 1947.

and then in the atom so excited being ionized through the absorption of a second quantum with energy greater than the ionization potential (3·38 eV.) of the $s = 2$ states. The second quantum is supplied by the radiation from the photosphere. By using (1) a cross-section $\sim 10^{-16}$ for the absorption of the second quantum, (2) the assumption that the photosphere radiates like a black body, it can be shown that the probability of this process is $\sim 2 \times 10^{-4}$ per absorption of a Lyman α-quantum. That is, a Lyman α-quantum would, on the average, experience ~ 5000 absorptions and re-emissions by neutral atoms before being lost by this ionization process.

We now show that, except near the top of the chromosphere, the rate of escape of Lyman α-quanta is small compared with the rate of production by process (*a*). As a consequence it follows, since Lyman α-quanta cannot accumulate indefinitely, that we must have

$$\left[\begin{array}{l}\text{The rate of production of Lyman } \alpha\text{-quanta through}\\ \text{electron-proton recombinations to the } s = 2 \text{ states}\\ \text{of the neutral hydrogen atom}\end{array}\right] \sim \left[\begin{array}{l}\text{The rate of loss of Lyman } \alpha\text{-quanta through ioni-}\\ \text{zation of the } s = 2 \text{ states due to absorption, from}\\ \text{radiation emitted by the photosphere, of quanta}\\ \text{with energies} > 3{\cdot}38 \text{ eV.}\end{array}\right] \qquad (20{\cdot}1)$$

In establishing the following conclusions we may ignore the deviations from ionization equilibrium, discussed in section 17, since it can be shown that such deviations lead to our conclusions applying *a fortiori*. On this basis it has been calculated from (7·10), (17·2), and (17·3) that the number of neutral atoms in a unit radial column containing *descending* material is

$$\sim 10^{13}(e^{x_1/kT} - 1) \text{ per cm.}^2, \qquad (20{\cdot}2)$$

where T is the temperature at the base of the unit column, and x_1 is the ionization potential ($2{\cdot}15 \times 10^{-11}$ ergs) of the $s = 1$ state of the hydrogen atom. The corresponding result for a unit radial column containing *ascending* material is

$$\sim 1{\cdot}5 \times 10^{13}(e^{x_1/kT} - 1) \text{ per cm.}^2. \qquad (20{\cdot}3)$$

The magnitudes of (20·2) and (20·3) for several values of T, T can be found with the aid of Table 22.

Table 22.

T or T (deg. K.)	2×10^4	5×10^4	10^5	2×10^5
$10(e^{x_1/kT} - 1)$ or $10(e^{x_1/kT} - 1)$	26,000	220	38	12

Now it is well known in diffusion theory that the number of collisions suffered by a particle in diffusing a distance d is $\sim(d/\text{mean-free path})^2$. A comparable result holds in the present problem. Thus, if we assume, for the moment, that the Lyman α-quanta succeed in escaping from the solar atmosphere, the number of absorptions and re-emissions experienced by a quantum is given to sufficient approximation by taking the square of the product of the absorption cross-section with either (20·2) or (20·3) (these expressions being of the same order of magnitude). For the temperatures occurring in the chromosphere, the absorption cross-section is $\sim 10^{-12}$. The estimates given by this procedure are therefore obtained by squaring the values in the second row of Table 22. For example, a Lyman α-quantum escaping from the solar atmosphere, that was initially emitted in material of temperature 10^5 deg. K., must have experienced ~ 1500 subsequent absorptions and re-emissions. Remembering that the probability of a Lyman α-quantum being lost through ionization is $\sim 2 \times 10^{-4}$ per absorption, it is seen that we have enclosure conditions for temperatures appreciably below 10^5 deg. K. Hence (20·1) can be applied except near the top of the chromosphere.

The rate of electron-proton recombinations, as given by (7·9), is insensitive to the kinetic temperature of the material. Accordingly, (20·1) remains valid if we replace the kinetic temperature of the chromosphere by the temperature of the photosphere. This modification of (20·1) shows the density n_2 of atoms in the $s = 2$ states must be given approximately by the thermodynamic condition

$$(20{\cdot}4) \qquad n_2 \sim 4 n_e n_p \left(\frac{h^2}{2\pi m_e k\theta}\right)^{3/2} e^{(x_1 - x_2)/k\theta},$$

where x_2 is the ionization potential (3·38 eV.) of the $s = 2$ states. Strictly, θ should be taken as the effective temperature of the photosphere. But since the interchanges among the states of principal quantum numbers 2, 3, 4, . . . , arise almost wholly from the absorption and re-emission of quanta supplied by the *reversing layer*, it is convenient to approximate by taking θ as the temperature of the reversing layer. Then the thermodynamic equation

$$(20\cdot5) \qquad n_s = s^2 n_e n_p \left(\frac{h^2}{2\pi m_e k\theta}\right)^{3/2} e^{(x_1 - x_s)/k\theta}$$

gives an order of magnitude estimate for the densities of atoms in all states with $s > 1$. The quantity x_s is the ionization potential in ergs of the states of principal quantum number s.

It is important to notice that (20·5) cannot be applied, in regions of high kinetic temperature, to the $s = 1$ state. Since $n_2, n_3, \ldots,$ are small compared with n_1, the ratio n_1/n_p is then given by (7·10), *which does not agree even approximately with the thermodynamic value corresponding to temperature* θ. The interchanges between the $s = 1$ and $s = 2$ states and between the $s = 1$ state and the continuum are determined essentially by the high kinetic temperature of material in the chromosphere, whereas the interchanges between states with $s > 1$ and between these states and the continuum are determined largely by the radiation emitted from the photosphere and the reversing layer. Accordingly, thermodynamic equations must apply approximately in the latter, but not in the former cases.

The present discussion differs from Wildt's analysis,† in that he assumes (20·5) to hold even when $s = 1$. This assumption, although not used in the main analysis, is responsible for Wildt's conclusion that the relative abundances of the metals are much lower in the chromosphere than at the photosphere. This result disagrees with our expectation that the convective stirring of the solar atmosphere produces a fairly uniform composition. Any non-uniformity from radiation pressure *increases* the relative abundances of metals in the chromosphere.

When we put $\theta = 4830°$ K., and $n_e = n_p$, equation (20·5) takes the form

$$(20\cdot6) \qquad \log n_s = \log s^2 + 1\cdot043(13\cdot53 - x_s) + 2 \log n_e - 20\cdot91,$$

where x_s is now expressed in eV. The quantity

$$\{\log s^2 + 1\cdot043(13\cdot53 - x_s)\}$$

is 4·13 when $s = 2$, and is given by Table 24 for $s > 2$. By using the known values of n_e (see Fig. 7) we obtain the estimates

† R. Wildt, *Ap.J.*, **105**, 36, 1947.

given in Table 23 for n_s. The number of $s = 2$ atoms lying in a unit column extending radially outwards from the base of the chromosphere $\sim 10^{13}$ per cm.^{2}, which gives a large opacity in $H\alpha$ and appreciable opacity in $H\beta$ and $H\gamma$. The corresponding number for a unit column extending along the line of sight from the earth and passing at a height of 1500 km. above the solar limb $\sim 10^{15}$ per cm.^{2}. This value leads to considerable self-absorption

Table 23.

Height above photosphere (km.)	1500	2500	4000	6000	8000
$\log n_e$	11·18	10·92	10·53	10·08	9·80
s					
2	5·58	5·06	4·28	3·38	2·82
3	3·97	3·45	2·67	1·77	1·21
4	3·54	3·02	2·24	1·34	0·78
5	3·41	2·89	2·11	1·21	0·65
6	3·40	2·88	2·10	1·20	0·64

in the higher members of the Balmer series. The self-absorption becomes less, however, for unit columns along the line of sight that pass through the chromosphere at increasing heights above the photosphere. Wildt has shown (*loc. cit.*) that self-absorption always becomes negligible at the greatest observed height of a line.

The rate of emission $E_{ss'}$ per unit volume in transitions from $s \to s'$ is given by

$$E_{ss'} = A_{ss'} h\nu_{ss'} n_s, \tag{20·7}$$

where $\nu_{ss'}$ is the frequency of the emitted quanta and $A_{ss'}$ are the Einstein coefficients for hydrogen. $E_{ss'}$ can be calculated for the Balmer series from (20·6) and (20·7) together with the values of $(A_{ss'} h\nu_{ss'})$ given in Table 24. But in general the values of $E_{ss'}$ so calculated are not readily compared with the observed intensities, on account of the self-absorption occurring in the lower regions of the chromosphere. To overcome this difficulty we shall employ an ingenious procedure due to Wildt (*loc. cit.*).

If there were no variation in detectability with frequency, the height of fade-out of every line would correspond to the same value of $E_{ss'}$ (remembering that there is no self-absorption at the greatest height attained by a line). But if a colour correction $\Delta C_{ss'}$, arising from the variation with frequency of photographic sensitivity, reflection loss in the optical system, and atmospheric extinction, is included, then at the height of fade-out every line has the same value of $(\log E_{ss'} + \Delta C_{ss'})$. Thus the height of fade-out of each member of the Balmer series satisfies the equation

$$(20\cdot8) \quad \text{constant} = \log E_{ss'} + \Delta C_{ss'} = \log(A_{ss'}h\nu_{ss'}) + \log s^2 + 1\cdot043(13\cdot53 - x_s) + 2\log n_e - 20\cdot91 + \Delta C_{ss'},$$

where the constant depends on the sensitivity of the observational equipment. The quantities $\log(A_{ss'}h\nu_{ss'})$ and

$$\{\log s^2 + 1\cdot043(13\cdot53 - x_s)\}$$

are given in Table 24 for $s' = 2$, $s = 3, 4, \ldots, 37$. To determine the constant in (20·8) we use the observed result that the line $H22$, corresponding to $s = 22$, $s' = 2$, attains a height ~2200 km., and that at this height $\log n_e \sim 11\cdot0$ (see Fig. 7). These values give

$$(20\cdot9) \qquad \text{constant} = 16\cdot57 + \Delta C_{22,\,2} - 20\cdot91,$$

and by using (20·8) we obtain

$$(20\cdot10) \quad 16\cdot57 + \Delta C_{22,\,2} = \log(A_{s2}h\nu_{s2}) + \log s^2 + 1\cdot043(13\cdot53 - x_s) + 2\log n_e + \Delta C_{s2}$$

for the Balmer series. Now, since $\log n_e$ is a known function of height, (20·10) is therefore an equation for the heights of fade-out of the lines of the Balmer series. With the exception of the cases $s = 3, 4, 5$ the theoretical values determined by (20·10) are shown in Table 24. The heights given by (20·10) for $H\alpha$, $H\beta$, and $H\gamma$ are > 10,000 km. These estimates have been rejected since the enclosure conditions for Lyman α-quanta are not satisfied near the top of the chromosphere. This breakdown of enclosure conditions must lead to a marked reduction in n_s, $s > 1$, and we therefore give 10,000 km. as the approximate height attained by these lines. Although the general agreement between the observed and the theoretical heights is on the whole satisfactory, there is a systematic difference for values of s between 7 and 18. It is not yet known whether this difference is due to imperfections in the theory or to errors of observations.

The values of ΔC_{s3} are rather uncertain. A similar discussion could otherwise be given for the Paschen series.

Finally, we note that chromospheric line emission arises essentially from the 'resonance scattering' of quanta present in the normal emission from the reversing layer and photosphere. That is, the observed intensities are not primarily due to the conversion of thermal energy to radiation. But although such a conversion of energy does not lead to an important contribution to the ob-

Table 24.

s	$\log(A_{s2}h\nu_{s2})$ erg./sec. as unit	$\log s^2 + 1{\cdot}043\,(13{\cdot}53 - x_s)$	$(\Delta C_{s2} - \Delta C_{22,2})_{obs}$	Height in km. as given by (20·10)	Observed height of fade-out (km.)
3	−3·88	+2·52	−·13	~ 10,000	12,000
4	4·47	2·09	+·58	~ 10,000	9000
5	4·94	1·96	·54	~ 10,000	8000
6	5·33	1·95	·36	7500	8000
7	5·67	1·98	·26	6000	8000
8	5·95	2·03	·19	4900	8000
9	6·21	2·08	·13	4200	7000
10	6·44	2·14	·10	4000	6000
11	6·64	2·20	·08	3500	6000
12	6·83	2·25	·06	3200	6000
13	7·01	2·31	·05	3000	5500
14	7·16	2·36	·04	2900	4500
15	7·31	2·41	·03	2700	4000
16	7·45	2·46	·02	2600	3500
17	7·59	2·51	·02	2500	3500
18	7·71	2·55	·01	2400	3000
19	7·83	2·60	+·01	2400	2500
20	7·94	2·63	·00	2300	2500
21	8·05	2·67	·00	2200	2500
22	8·14	2·71	·00	2200	2200
23	8·24	2·75	·00	2100	2000
24	8·33	2·78	·00	2000	2000
25	8·42	2·82	·00	1900	1800
26	8·51	2·85	·00	1800	1800
27	8·59	2·88	·00	1700	1500
28	8·67	2·91	·00	1600	1500
29	8·75	2·94	·00	1500	1200
30	8·82	2·97	·00	1400	1200
31	8·89	2·99	·00	1300	1000
32	8·96	3·02	·00	1250	1000
33	9·03	3·05	·00	1200	1000
34	9·09	3·07	·00	1100	800
35	9·15	3·10	·00	1000	800
36	9·21	3·12	·00	950	600
37	−9·27	+3·15	·00	900	500

served line radiation from the chromosphere, the emission and absorption of Lyman α-quanta, due to the high kinetic temperature of chromospheric material, is important in increasing n_s, $(s > 1)$, and thereby in increasing the scattering efficiency of the hydrogen.

CHAPTER V

ELECTROMAGNETISM IN SOLAR PHYSICS

21. Prominences

A prominence is a localized region in the corona where the temperature is low enough for many of the normal chromospheric lines to be visible. According to section 17 the normal rate of radiation by the corona is about 1% of the rate at which energy is supplied by accretion. Thus it follows, since the rate of radiation depends on the square of the density, that a tenfold increase of density is required to produce an appreciable degree of cooling. Moreover, since this increase of density must occur *before* radiation becomes important, the hydrostatic pressure must also increase by a factor ~ 10. Such an increase in pressure can arise from the magnetic focusing of material.

It was seen in section 5 (iv) that the magnetic field due to the electric currents in a given domain cannot produce an appreciable effect on the material in a second separate domain when, as in the corona, the condition (5·11) is satisfied. Thus the distant action of the currents in a sunspot or in any other interior region of the sun cannot appreciably affect the motion of material in the corona. The requirement for magnetic effects to become important can be understood from (5·2). The distant action between electric currents in a given domain and material in a second separate domain involves only the small $\boldsymbol{H} \times \partial \boldsymbol{E}/\partial t$ term. To produce large effects the $\boldsymbol{H} \times \operatorname{curl} \boldsymbol{H}$ term must be brought into operation, and by (5·1) this requires electric currents to flow in the corona. That is, in order for important magnetic processes to arise, electric currents of appreciable magnitude must be propagated from the solar interior into the corona. Assuming that this condition is satisfied, we now examine the cooling requirements discussed in the previous paragraph.

The necessary degree of magnetic focusing cannot be obtained unless the $\boldsymbol{H} \times \operatorname{curl} \boldsymbol{H}$ term in (5·2) exceeds the grad P term by a factor ~ 10 at least. Taking $|\operatorname{curl} \boldsymbol{H}|$ to be of the same order of magnitude as the space derivatives of $\boldsymbol{H}$ (we assume here that sufficiently large currents have been propagated into the corona) this condition gives

(21·1) $$H^2/8\pi > \sim 10P.$$

Using (16·10) this becomes

(21·2) $$|\boldsymbol{H}| > \sim 3{\cdot}1\,(h^*/h)^{\frac{1}{2}} \text{ gauss},$$

which is satisfied everywhere by an appreciable margin since the general magnetic field of the sun ~ 50 gauss. Hence, the propagation into the corona of electric currents from the interior of the sun is the essential factor in the origin of prominences. In order for $|\operatorname{curl} \boldsymbol{H}|$ to be of the same order as the space derivatives of $\boldsymbol{H}$ the electric currents in the corona and the currents in the solar interior must be of the same order of magnitude.

Since (21·2) is always satisfied in the corona we reach the important conclusion that prominences may occur anywhere on the solar disk. The places on the solar surface from which the efflux of electric current occurs represent the centres of attraction described in section 9 (ii). It is clear that the positions of prominences are determined by the centres of attraction and not by any special property of the material in the prominences themselves. The diversity in type is attributed to the complex distributions of electric current and magnetic field that can arise through intricate convection processes.

The rate of radiation by ionized hydrogen increases with decreasing temperature. Thus it follows that, once cooling in a localized region is started, the radiating process becomes increasingly strong until the temperature falls to a value $\sim 10{,}000°$ K. At this stage the rate of radiation is reduced by a rapid decrease in the proportion of ionized hydrogen. Accordingly, we expect the temperatures in prominences to be $\sim 10{,}000°$ K. In order to maintain a constant hydrostatic pressure during the cooling process there must be an inflow of material from surrounding regions. If the pressure is maintained at $10P$, where P is given by (16·10), the density is $\sim 5 \times 10^{11}$ atoms/cm.^{3} when the temperature is $\sim 10{,}000°$ K. At this large density the cooled material falls under gravity along the magnetic lines of force, and is observed as streamers that enter the centres of attraction. The velocity attained by material falling from height h is $\sim(gh)^{\frac{1}{2}}$. If the prominence is to be quasi-stable the falling material must be replaced by the influx of new material from surrounding regions of the corona. That is, there must be a continuous condensation of coronal material. The balance between the new inflowing material and the cooled material falling under gravity is a critical feature in establishing the quasi-stability of a prominence; it can be shown that the prominence must be of the characteristic filament form in order that this balance be achieved.

Finally, we note that the dark appearance (in $H\alpha$, for example) of a prominence, when projected on the solar disk, is not explained by the argument that there is a nett conversion of radiation into thermal energy taking place within the prominence, as a result of the temperature being lower than in the reversing layer. This suggestion is inadequate because it is based on strictly thermodynamic concepts. A proper solution of this difficult problem is beyond the range of the present work.

22. The Magnetic Cycle

Both the variations in the frequency and distribution of prominences (see section 9 (ii)) and the changes in the shape of the outer corona (see section 9 (i)) indicate that the general magnetic field of the sun does not remain constant during the solar cycle. Moreover, the reversal at sunspot minimum of the magnetic polarities of the leader spots in bipolar groups provides direct observational evidence in favour of a magnetic cycle (see section 3 (v)).

The changes in the shape of the outer corona probably arise through a process similar to the formation of prominences. That is, electric currents are propagated into the outer corona and the $\boldsymbol{H} \times \text{curl}\, \boldsymbol{H}$ term in (5·2) is brought into operation. In opposition to this view it is sometimes argued that the changes of shape arise from the action of the magnetic fields of sunspots. But there are important objections to this suggestion. For example, cyclic changes of shape occur in polar regions where there are no sunspots. Moreover, if the general magnetic field is taken as ~ 50 gauss at the photosphere, the magnetic moment corresponding to the general field is considerably greater than the magnetic moment of a large sunspot group. Thus for $r > 2$ the general field is appreciably larger than the field due to a sunspot group.

In anticipation of the work of Appendix II we assume that the magnetization of the sun is confined essentially to the surface layer defined by $0{\cdot}85 < r < 1$. The form of the magnetic field in this layer can change with time due to a convective circulation. An example is shown, for a typical meridian plane, in Fig. 8, where for illustrative purposes we neglect the curvature of the

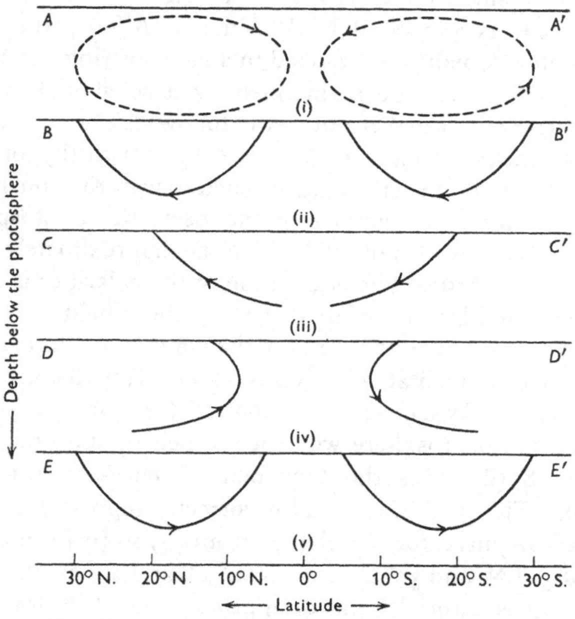

Fig. 8. (i) shows the convective circulation; (ii), (iii), (iv), (v) represent successive positions of the regions of high magnetic intensity.

photosphere which is represented by AA', BB', CC', DD', and EE' in the respective subdivisions of the figure. Typical stream-lines of an assumed convective circulation in meridian planes are shown in (i). The shapes of regions of high magnetic energy are altered continuously by this convection. It is useful in this connexion to think of a region of high magnetic energy as a flexible magnet. The initial position of such regions are indicated by the lines shown in (ii). The lines in (iii), (iv) give subsequent positions of these regions, and (v) represents the position after a *half-cycle* of the convection currents. The arrows in (ii), (iii), (iv), and (v) show the directions of the magnetic lines of force. As in the case of an ordinary magnet, the lines of force extend outside the regions of high magnetic energy and form closed circuits. This is not shown in Fig. 8.

The magnetic cycle discussed in the previous paragraph refers to the general field of the sun. Sunspots are associated with comparatively small domains carried by the general field. It is probable that, as suggested by Walén and Alfén,† the lines of force in such a domain are arranged in an anchor ring. A magnetic field of this form can be maintained by a solenoidal system of electric currents. Furthermore, we follow Alfén in supposing that these sunspot magnetic rings are systematically orientated with respect to the general field in such a way that on rising to the solar surface a magnetic ring intersects the photosphere in two areas of opposite polarity which lie approximately on the same circle of latitude. The orientation of the polarities is assumed to be determined by the polarity of the general field.

By taking a period ~ 22 years for the convective circulation, we can now give a qualitative description of many features of the sunspot cycle. Whenever a region of high magnetic energy approaches the photosphere we expect a belt of sunspots to arise. Thus Fig. 8 (ii) gives the four belts associated with sunspot minimum. The polarities are also correctly represented. In the $\sim 3{\cdot}7$ years required for the change from (ii) to (iii) the two belts initially at 30° N. and 30° S. drift to 20° N. and 20° S. respectively, and the two equatorial belts disappear. The drift towards the equator continues during the $\sim 3{\cdot}7$ years required for the change from (iii) to (iv). In the change from (iv) to (v), regions of high magnetic energy approach the photosphere at 30° N. and 30° S., thereby leading to the outbreak of new sunspot cycle. The belts of sunspots corresponding to (v) occupy positions similar to (ii).

It will be seen from the directions of the arrows that the polarities in (v) are opposite to the polarities in (ii). It follows therefore that the form of circulation shown in Fig. 8 gives a natural explanation of the reversal of the magnetic polarities. A further half-cycle of the convection currents would bring (v) back to the polarities shown in (ii). Thus the period of the magnetic

† C. Walén, *Ark. f. mat. astr. och. fysik*, Bd. **30**A, No. 15, and Bd. **31**B, No. 3, 1944. H. Alfén, *M.N.*, **105**, 3 and 382, 1945.

cycle is equal to the period of the convection currents. The present discussion differs from the views advocated by Alfén (*loc. cit.*), who takes the field of a static dipole to represent the general field of the sun. This assumption of necessity leads to results that do not show the required reversal of polarities at sunspot minimum.

Provided there is no general disintegration of the material within the sunspot magnetic rings, then, on the short-term view usually adopted in solar physics, these rings can be regarded as having a permanent existence. It may be noted in this connexion that the work of section 5 (iii), which gave ~ 300 years for the lifetime of the magnetic field of a sunspot, probably underestimates the decay time. Thus, in the first place, we took the spot radius ($\sim 10^9$ cm.) as the characteristic dimension a of the field, whereas a better value is obtained by using the inner radius ($\sim 3 \times 10^9$ cm.) of the magnetic ring. Furthermore, we assumed a conductivity corresponding to the photospheric layers ($\sim 10^{-8}c^2$), whereas it is probable that the sunspot magnetic rings spend the main portion of their lifetimes at deeper levels where $\sigma'/c^2 \sim 10^{-6}$. Accordingly $\sigma' a^2/c^2 \sim 300{,}000$ years represents a better estimate for the decay time.

The above remarks are concerned with a magnetic cycle occurring in the equatorial region between 30° N. and 30° S. It may be noted, however, that the distribution and periodicity of prominences in higher latitudes suggests that similar magnetic cycles occur in polar regions. The absence of sunspots in high latitudes requires the sunspot magnetic rings to be absent from these polar circulations.

Finally, the present work may be considered in relation to a recent paper by Blackett,† who has pointed out that the magnetic moments of the earth, the sun, and the star 78 Vir are given approximately by the formula

$$\beta G^{\frac{1}{2}} \boldsymbol{h}/c, \tag{22·1}$$

where $\boldsymbol{h}$ is the angular momentum of the body under consideration, and β is a dimensionless constant of order unity. Blackett has inferred from this remarkable observation that all massive rotating bodies possess magnetic moments given by (22·1), and he suggests

† P. M. S. Blackett, *Nature*, **159**, 658, 1947.

this result does not arise from specific processes, but represents a fundamental property of matter. The above investigation does not support this suggestion. Thus, the necessity for a magnetic cycle requires the action of specific processes. Further evidence for the importance of specific processes follows from the discussion, given in Appendix II, of the origin of the general magnetic field of the sun.

23. Electric Fields

(i) *In a sunspot*

As pointed out at the beginning of section 5 (iii), the transverse conductivity σ'' may be neglected in the main body of a sunspot. Then we have the following relation between the magnitudes of the current density, and the electric field as measured by an observer moving with the material:

$$(23{\cdot}1)\qquad |\boldsymbol{E}| \sim k\,|\boldsymbol{j}|/c^2T^{3/2},$$

where $k \sim 6{\cdot}8 \times 10^{13}$. Now the average value, $\overline{|\boldsymbol{H}_s|}$, say, of the magnetic field in a sunspot is given, so far as order of magnitude is concerned, by

$$(23{\cdot}2)\qquad \overline{|\boldsymbol{H}_s|} \sim 4\pi a\,\overline{|\boldsymbol{j}|}/c,$$

where a is the characteristic dimension of the field, and $\overline{|\boldsymbol{j}|}$ is an average value of the current density. The corresponding average value of the electric field is

$$(23{\cdot}3)\qquad \overline{|\boldsymbol{E}|} \sim k\,\overline{|\boldsymbol{H}_s|}/4\pi acT^{3/2}.$$

By putting $k = 6{\cdot}8 \times 10^{13}$, $\overline{|\boldsymbol{H}_s|} = 3000$ gauss, $a = 3 \times 10^9$ cm. in (23·3) we obtain the values given in Table 25.

Table 25.

T (deg. K.)	5×10^3	2×10^4	5×10^4	10^5	10^6
$\overline{\lvert\boldsymbol{E}\rvert}$ (10^{-8} volt per cm. as unit)	15	1·9	·48	·17	·0054

Except at points where a local focusing of the electric current occurs, the electric field, as recognized by an observer moving with the conducting material, cannot exceed the order of magnitude given by (23·3). But an observer rotating with the sun and situated at a definite point of the solar surface may measure a substantially higher field. Thus the induced electric field, given by applying Faraday's law to the change of magnetic field shown in Fig. 1, is $\sim 0 \cdot 1$ volts/cm. The large difference between this value and (23·3) shows that the material in a sunspot does not remain fixed during the growth of the magnetic field (the electric currents would otherwise be impossibly strong), but that the material and the magnetic field move substantially together. An estimate for the magnitude of the velocity of material in a sunspot relative to surrounding material is given by equating the induced field $\sim 0 \cdot 1$ volts/cm., measured by an observer rotating with the sun, to

$$|\boldsymbol{H}_s| \,.\, (\text{relative speed})/c.$$

For $|\boldsymbol{H}_s| = 3000$ gauss this gives a relative speed $\sim 3 \times 10^3$ cm./sec.

So far we have neglected the general magnetic field $\boldsymbol{H}_g$. Interesting effects occur when a relative motion $\boldsymbol{u}$ exists between the magnetic field $\boldsymbol{H}_g$ and the magnetic field $\boldsymbol{H}_s$ of a sunspot. We note that since $\boldsymbol{H}_g$ may be taken as independent of time over a period ~ 10 days, the magnitude of $\boldsymbol{u}$ must be $\sim 3 \times 10^3$ cm./sec. during the observed growth of the magnetic field of the sunspot. The electric current in the spot arises from a field

$$(23 \cdot 4) \qquad (\boldsymbol{u}_s \times \boldsymbol{H}_s + \boldsymbol{u}_g \times \boldsymbol{H}_g)/c,$$

where $\boldsymbol{u}_s$, $\boldsymbol{u}_g$ represent the velocities of material relative to the fields $\boldsymbol{H}_s$, $\boldsymbol{H}_g$ respectively. The discussion of the previous paragraph gives $|\boldsymbol{u}_g| \sim 3 \times 10^3$ cm./sec. Accordingly, putting $|\boldsymbol{H}_g| \sim 50$ gauss, it follows that $|\boldsymbol{u}_g \times \boldsymbol{H}_g|/c$ is in general $\sim 1 \cdot 5 \times 10^{-3}$ volts/cm. With the exception of a special case discussed below, (23·4) must be of the same order of magnitude as (23·3). Thus, since (23·3) is very small compared with $1 \cdot 5 \times 10^{-3}$ volts/cm., the two terms in (23·4) must, except in this special case, nearly cancel. That is

$$(23 \cdot 5) \qquad \boldsymbol{u}_s \times \boldsymbol{H}_s + \boldsymbol{u}_g \times \boldsymbol{H}_g \sim 0.$$

From the definitions of $\boldsymbol{u}_g$, $\boldsymbol{u}_s$, $\boldsymbol{u}$ we have $\boldsymbol{u}_g = \boldsymbol{u}_s - \boldsymbol{u}$. Hence (23·5) can be written as

$$(23 \cdot 6) \qquad \boldsymbol{u}_s \times (\boldsymbol{H}_s + \boldsymbol{H}_g) \sim \boldsymbol{u} \times \boldsymbol{H}_g.$$

It is easy to show from (23·6) that provided $(\boldsymbol{H}_s + \boldsymbol{H}_g) \neq 0$, $\boldsymbol{u}_s$ must be of the form

$$(23{\cdot}7) \quad \{(\boldsymbol{H}_s + \boldsymbol{H}_g) \times (\boldsymbol{u} \times \boldsymbol{H}_g) + \lambda(\boldsymbol{H}_s + \boldsymbol{H}_g)\}/|\boldsymbol{H}_s + \boldsymbol{H}_g|^2,$$

where λ is an arbitrary constant. This result may be regarded as determining $\boldsymbol{u}_s$ when $\boldsymbol{H}_g$, $\boldsymbol{H}_s$ and $\boldsymbol{u}$ are given. In the main body of a sunspot $|\boldsymbol{H}_g| \ll |\boldsymbol{H}_s|$ and λ can be chosen so that $|\boldsymbol{u}_s| \ll |\boldsymbol{u}|$. On the other hand, in regions surrounding a sunspot, where $|\boldsymbol{H}_s|$ is comparable with $|\boldsymbol{H}_g|$, the quantity $|\boldsymbol{u}_s|$ is in general $\sim|\boldsymbol{u}|$.

The solution (23·7) breaks down in the immediate neighbourhood of a neutral point, where $(\boldsymbol{H}_s + \boldsymbol{H}_g) = 0$ to a high degree of approximation. In this case (23·5) gives

$$(23{\cdot}8) \qquad \boldsymbol{u}_s = \boldsymbol{u}_g + \lambda \boldsymbol{H}_s,$$

where λ is again arbitrary. Hence, if (23·5) is satisfied at and near a neutral point of the magnetic field it follows that $\boldsymbol{u}$ must be parallel to both $\boldsymbol{H}_s$ and $\boldsymbol{H}_g$. That is, in the neighbourhood of a neutral point there would be no relative motion between the lines of force of the field $\boldsymbol{H}_s$ and the lines of force of the general field $\boldsymbol{H}_g$. Since $\boldsymbol{H}_g$ may be taken as constant for periods $\sim$10 days, it is seen that $\boldsymbol{H}_s$ would have to remain constant at a neutral point during the growth of the magnetic field of a sunspot. This conclusion is in conflict with the indications of observation. It seems probable therefore that (23·5) is not even approximately satisfied at a neutral point. This is possible because the discussion leading to (23·5) was based on average values that need not apply at a particular point. If (23·5) is not approximately satisfied, the electric field recognized by an observer moving with the material is $\sim 10^{-3}$ volts/cm. Thus, since the corresponding current density must exceed by a large factor the mean value for the interior of a sunspot, it follows that the focusing of current at and near a neutral point must be particularly strong. A similar conclusion has been reached by Giovanelli,† who uses arguments different from those given above.

(ii) *In the solar atmosphere*

When the magnetic field is zero the transverse conductivity σ^{II} is also zero. Accordingly, the conclusions reached above

† R. G. Giovanelli, *M.N.*, **107**, 338, 1947, and other papers in the press.

concerning the focusing of current at a neutral point of the magnetic field apply also in the solar atmosphere. But at points in the chromosphere and corona where $|\boldsymbol{H}_s| \gg |\boldsymbol{H}_g|$ the transverse conductivity is important, and the limitation in the electric field (measured by an observer moving with the material), as given by Table 25, is confined to the component of $\boldsymbol{E}$ parallel to $(\boldsymbol{H}_g + \boldsymbol{H}_s)$. The components of $\boldsymbol{E}$ perpendicular to $(\boldsymbol{H}_g + \boldsymbol{H}_s)$ may appreciably exceed the values given in this table.

24. Solar Flares †

(i) *The condition for a discharge*

The formulæ given in section 5 (iii) for the electrical conductivity of ionized hydrogen depend on the assumption that the electrical energy, acquired by an electron in moving through a displacement $(Q_{ep}n_p)^{-1}$ directed along the magnetic field, may be regarded as negligible. The quantity Q_{ep} is the electron-proton collision cross-section given by

$$(24{\cdot}1) \qquad \begin{aligned} Q_{ep} &= X/\eta^2, \\ X &= 2\pi\epsilon^4 \ln\{10^{-3}\eta/\epsilon^2\}, \end{aligned}$$

where η is the electron energy relative to the proton. This condition can be regarded as satisfied except under the special circumstances examined below.

There is an appreciable probability that an electron of initial energy η_0, moving in an electrical field $\boldsymbol{E}$, will be continuously accelerated if the energy acquired in a displacement $\eta_0^2/n_pX(\eta_0)$ along the magnetic field is $\sim\eta_0$. This continuous acceleration is possible because, on account of the rapid decrease of Q_{ep} with increasing η, the direction of motion of the electron is no longer randomized by collisions with the protons. Expressed quantitatively, the condition for the occurrence of 'runaway' electrons is

$$(24{\cdot}2) \qquad \epsilon\eta_0^2 E_m/n_pX(\eta_0) \sim \eta_0,$$

where E_m is the component of $\boldsymbol{E}$ along the magnetic field. Equation (24·2) may be regarded as determining η_0 when E_m is given.

† Although the present theory differs in form from that given by Giovanelli, it is emphasized that this investigation was prompted by Giovanelli's important papers.

Since $X(\eta)$ varies slowly with η it is sufficiently accurate to replace η_0 in $X(\eta_0)$ by kT. Thus

$$(24\cdot3) \qquad \eta_0 \sim X(kT)n_p/\epsilon E_m,$$

and the proportion of runaway electrons in ionized hydrogen at temperature T is

$$(24\cdot4) \qquad \sim \exp\left[-\frac{\eta_0}{kT}\right] \sim \exp\left[-\frac{n_p X(kT)}{\epsilon kTE_m}\right]$$

Except in the special case discussed in the present section this fraction is negligibly small.

A selection of values of $n_p X(kT)/\epsilon kT$, for material in descending columns, is given in Table 26. The corresponding values for

Table 26.

Height above photosphere (km.)	1500	2500	4000	6000	8000
T (deg. K.)	5000	40,000	80,000	130,000	180,000
$\log n_p$	11·18	10·92	10·53	10·08	9·80
$\frac{n_p X(kT)}{\epsilon kT}$ (10^{-8} volt per cm. as unit)	$3\cdot64\times10^7$	$3\cdot15\times10^6$	$6\cdot85\times10^5$	$1\cdot56\times10^5$	$6\cdot10\times10^4$

material in the ascending columns exceed these results by a factor of order 5. It is seen from a comparison of Tables 25 and 26 that at normal points of the solar atmosphere the proportion of runaway electrons is quite negligible. At a neutral point of the magnetic field, on the other hand, E_m may be as large as 10^{-3} volts/cm. Accordingly, it is near such points that we must seek conditions suitable for a discharge. Indeed, the dependence given in Table 26 of $n_p X(kT)/\epsilon kT$ on height above the photosphere shows that, if a neutral point rises to a height appreciably greater than 4000 km., the proportion of runaway electrons becomes of order unity. The following work suggests that a flare occurs whenever this condition is satisfied. The rapid onset of a flare is due to the critical dependence of (24·4) on height.

(ii) *The development of a flare*

In the immediate neighbourhood of a neutral point of the magnetic field we expect the electric field to be $\sim 10^{-3}$ volts/cm.

Although it is difficult, in the present preliminary stages of the theory of solar flares, to give a precise determination of the size of such a region, we may take 100 km. as a rough estimate for the characteristic dimension. A runaway electron accelerated through this distance acquires an energy $\sim$10,000 eV. After leaving the vicinity of a neutral point the electron must follow a line of magnetic force until its energy is lost by collisions with the hydrogen present in the chromosphere and reversing layer. The majority of the magnetic lines of force enter and leave the solar atmosphere at the photosphere. Thus the bulk of the runaway electrons must eventually move downwards towards the photosphere. The depth of penetration into the reversing layer evidently depends on the stopping-power of hydrogen to fast electrons. Values given by Mott and Massey † show that to stop a 10,000 volt electron $\sim 10^{20}$ hydrogen atoms/cm.^{2} are required. Remembering that the scale height in the reversing layer $\sim$146 km. it follows therefore that the runaway electrons penetrate to a depth where the hydrogen density $\sim 10^{13}$ atoms/cm.^{3}. This density occurs at a height of about 1000 km. above the photosphere (see Fig. 7).

The above remarks suggest that the emission observed as a flare occurs, not at a neutral point, but in a region with a depth of the order of the scale height in the reversing layer, and in which the hydrogen density $\sim 10^{13}$ atoms/cm.^{3}. The area of the emitting region, seen in projection against the photosphere, is comparable with the size of the associated spot group. Accordingly, a flare has a much greater area than a projection of the small accelerating region near the neutral point. This arises because the electrons are guided along magnetic lines of force that fan out after leaving the accelerating region.

When a state of quasi-stability is reached the rate of emission by the whole flare must equal the rate at which the accelerating electrons gain energy from the electric field. Now it can be shown that, provided the neutral point attains a height such that (24·4) is of order unity, the rate at which the electric field does work is $\sim n_e \epsilon^{3/2} b^{7/2} |\boldsymbol{E}|/m_e^{\frac{1}{2}}$, where b is the characteristic dimension of the accelerating region, and $\boldsymbol{E}$, n_e are the electric field and the

† N. F. Mott and H. S. W. Massey, *The Theory of Atomic Collisions*, Oxford, 1933, p. 183.

electron density respectively at the neutral point. Thus the average emission is

$$(24\cdot5)\qquad \sim n_e \epsilon^{3/2} b^{7/2} \,|\, \boldsymbol{E} \,|\, / m_e^{\frac{1}{2}} A \text{ per unit area per unit time,}$$

where A is the area of the flare. As an example, we put $n_e = 10^{10}$ electrons/cm.^{3}, $b = 100$ km., $|\boldsymbol{E}| = 10^{-3}$ volts/cm., and $A = 10^{20}$ cm.^{2}. These values give an emission $\sim 3\cdot6 \times 10^8$ ergs/cm.^{2} per sec. when the flare is projected against the photosphere. This result is of the required order of magnitude.

(iii) *The effective width of the Hα-line*

The hydrogen of the reversing layer exposed to collisions with runaway electrons must become largely ionized. For otherwise there would be an enormous ionization rate. The energy loss per ionization,† for a 10,000 volt electron moving in neutral hydrogen, is ~ 60 eV. Thus, taking $3\cdot6 \times 10^8$ ergs per cm.^{2} per sec. as the energy carried by the runaway electrons, it follows that if the hydrogen in the reversing layer were neutral the ion production would be $\sim 3\cdot8 \times 10^{18}$ per cm.^{2} per sec. With this rate, the hydrogen exposed to collisions would, if initially neutral, become largely ionized in ~ 100 sec. Accordingly, we must take the electron density in the main emitting region of a flare to be $\sim 10^{13}$ per cm.^{3}.

This electron density is large enough to provide an explanation, on the basis of the Stark effect, of the large effective widths of the hydrogen emission lines in flare spectra.‡ (We note that these great widths cannot be explained on the basis of Doppler broadening, since this would involve a temperature $\sim 10^6$ deg. K. for the emitting material. The normal metallic lines observed in flare spectra, such as the *H*- and *K*-lines of *Ca* II, evidently could not be emitted at such a temperature.) Employing a formula due to Holtzmark,§ and inserting the constants appropriate to $H\alpha$, we have

$$(24\cdot6)\qquad k(\Delta\lambda) = 3\cdot13 \times 10^{-16} n_e (2\cdot61\epsilon)^{3/2} / (\Delta\lambda)^{5/2},$$

where $k(\Delta\lambda)$ is the absorption coefficient at distance $\Delta\lambda$ (Ang-

† N. F. Mott and H. S. W. Massey, *Op. cit.*, p. 183.
‡ M. A. Ellison and F. Hoyle, *The Observatory*, **67**, 181, 1947.
§ Cf. A. Unsöld, *Physik d. Stern Atmosphären*, J. Springer, 1938, p. 183.

ströms) from the line centre. The effective half-width is given approximately by solving the equation

$$(24{\cdot}7) \qquad 5n_2k(\Delta\lambda)(\mathfrak{R}\theta/g) = 1$$

for $\Delta\lambda$, where θ is the normal temperature of the reversing layer. The quantity n_2 represents the density of hydrogen atoms in the $s = 2$ states. On account of the large rate of production of Lyman α-quanta in the main emitting region, the value of n_2 must be calculated in terms of the electron density by the method of section 20. In particular, (20·6) gives $n_2 \sim 1{\cdot}66 \times 10^{-17}n_e^2$. Using this value, together with (24·6), (24·7), we obtain

$$(24{\cdot}8) \qquad \Delta\lambda \sim 7{\cdot}7 \times 10^{-16}n_e^{6/5}.$$

Hence, by putting $n_e \sim 10^{13}$ per cm.^{3} we have $\Delta\lambda \sim 3{\cdot}1A$ for the half-width of $H\alpha$. This result is in good agreement with the observed half-widths in typical flares.

(iv) *Prominence flares*

It was assumed throughout the above investigation that motion through the corona and chromosphere has little effect on the velocities of the runaway electrons. At the normal densities prevailing in the solar atmosphere this assumption is valid, but an exceptional case arises if the runaway electrons enter a prominence. In section 21 the hydrogen density in a prominence was estimated as $\sim 5 \times 10^{11}$ atoms/cm.^{3}. Thus, taking 10^{20} hydrogen atoms/cm.^{2} as sufficient to stop the high-speed electrons, it is seen that the depth of penetration into a prominence cannot exceed $\sim$2000 km. It follows therefore that runaway electrons moving along magnetic lines of force that enter a prominence are stopped in the outer fringes of the prominence. The emission from such a region is in many respects similar to that occurring in flares situated in the reversing layer. But the lower electron density occurring in a prominence flare ($\sim 5 \times 10^{11}$ per cm.^{3} as compared with $\sim 10^{13}$ per cm.^{3} in a normal flare) produces an important difference. Thus an electron density $\sim 5 \times 10^{11}$ per cm.^{3} is too low to give appreciable Stark broadening in the hydrogen lines. Accordingly, the line breadths occurring in prominence flares are probably due to the mass motions of the emitting material.

Prominence flares have recently been observed by d'Azambuja and by Ellison. †

25. Double-Stream Motion in a Magnetic Field

So far, we have considered, in the presence of a magnetic field, only a single stream of particles. We shall now discuss the interpenetration of two electrically neutral uniform streams of ionized material, one moving rectilinearly with velocity $\boldsymbol{u}_1$ and the other with rectilinear velocity $\boldsymbol{u}_2$. Furthermore, we suppose the cross-sections for collisions between the particles to be so small that, in the absence of a magnetic field, there is effectively free penetration between the two streams.

Interesting effects occur in the presence of a magnetic field $\boldsymbol{H}$. It was seen in section 5 (iv) that, for a single stream, the lines of magnetic force move with the stream, otherwise very large forces would quickly destroy the motion. In the two stream case the lines cannot move with both streams except in the special case where $(\boldsymbol{u}_2 - \boldsymbol{u}_1)$ is parallel to $\boldsymbol{H}$. In the general case ‡ electromagnetic forces must operate so as to destroy the component of $(\boldsymbol{u}_2 - \boldsymbol{u}_1)$ perpendicular to $\boldsymbol{H}$.

The above remarks have application to the motion of material in region 2. We see that a further complication is added to this motion, in that the relative velocity between the incoming accreted atoms and the material of the solar atmosphere must be parallel to the magnetic field. In general, the magnetic field possesses an appreciable vertical component and when this is the case the analysis of chapter III is not seriously affected by this modification. On the other hand, in exceptional parts of the solar atmosphere, the magnetic field may be strictly perpendicular to the vertical direction. The discussion of the properties of region 2 would then be more difficult than the work of region 2.

† The writer is indebted to Dr. Ellison for a valuable discussion on this question.

‡ This effect would form the subject of an interesting laboratory experiment.

CHAPTER VI

SOLAR AND TERRESTRIAL RELATIONSHIPS

26. Corpuscular Emission from the Sun

(i) *Preliminary remarks*

The permitted motion between particles ejected from the sun and the incoming accreted atoms is along the magnetic field. This relative motion cannot lead to the ejected particles escaping from the sun, since the magnetic lines of force form closed loops. It follows therefore that outward-moving particles cannot escape completely from the sun unless, over some particular area of the solar surface, the rate at which they carry momentum exceeds the rate at which inward momentum is supplied by the accreted hydrogen. In this case accreted atoms are prevented from reaching the area in question. But if the opposite situation occurs (in which the outward-moving particles carry less momentum than the accreted atoms) the ejected particles are swept back into the sun by the accreted material. We see therefore that a critical requirement must be met in order that particles ejected by the sun may reach the Earth. The following work of this section shows that this condition can be met only over localized areas of the solar surface.

(ii) *The expulsion of calcium by radiation pressure*

Although under normal circumstances the radiation pressure acting on certain atoms (see Table 21) may exceed the downward force due to gravity, the condition discussed in (i) is not satisfied. During solar flares, however, there is a sharp increase in the radiation pressure acting on the *Ca* II atoms present in the affected region. The question arises as to whether the outward momentum, carried by the radiation in the *H*- and *K*-lines, is sufficient to satisfy the requirement of (i). The energy emission in these lines is $\sim 3 \times 10^8$ ergs per cm.^{2} per sec. and this corresponds to an outward pressure of $3 \times 10^8/c$ dynes/cm.^{2} This value must be compared with the normal rate $m_p q g N(r^*)$ at which momentum is added by the accreted hydrogen atoms. Putting $g = 2{\cdot}1 \times 10^4$ cm. sec.$^{-1}$, $N(r^*) = 1{\cdot}36 \times 10^{18}$ atoms/cm.^{2}, we see that

$$3 \times 10^8 > m_p qcgN(r^*)$$

if $q = 0{\cdot}21$. This result shows that the intense radiation in the *H*- and *K*-lines of *Ca* II, occurring during a solar flare, is of the order of magnitude required to expel a beam of calcium atoms in opposition to the incoming accreted material. According to the work of section 25, no accreted material can reach the affected area during the expulsion of such a beam.

If we assume that the outward momentum carried by a beam of ejected calcium atoms is equal to the rate at which momentum is emitted in the *H*- and *K*-lines, then the rate at which the calcium atoms carry momentum across a sphere of radius r (radius of photosphere as unit) is $\sim 3 \times 10^8/cr^2$ dynes/cm.^{2}. The density of calcium atoms is then given, as a function of r, on dividing this expression by twice the mean kinetic energy of the ejected atoms. Now the velocity of the calcium atoms may be taken as ~1000 km./sec., in accordance with a well-known estimate due to Milne. (The rate of radiation under flare conditions is greater than that assumed by Milne, and on account of this Milne's value for the velocity of ejection might appear to be too small. But the effect of increased radiation is offset by the circumstance that radiation pressure is only effective while the calcium atoms are singly ionized. In passing through the corona an appreciable fraction of the ejected atoms must become multiply ionized. Accordingly, radiation pressure is in general effective only during the early stages in the outward motion of the calcium atoms.) With this velocity the density of calcium atoms is $\sim 1{\cdot}5 \times 10^4/r^2$. At the Earth r has a mean value ~215, and the corresponding density is ~0·33 calcium atoms/cm.^{3}. The total number of atoms in a column of unit cross-section lying between the sun and the Earth is $\sim 4{\cdot}9 \times 10^{12}$ per cm.^{2}. If these atoms were all singly ionized the absorption in the *H*- and *K*-lines would greatly exceed the absorption observed by Richardson † and by Brück and Rutllant.‡ This indicates that the majority of the calcium atoms become multiply ionized in passing through the corona.

† R. S. Richardson, *Ann. Report Mt. Wilson Obs.*, 1943–44.
‡ H. A. Brück and F. Rutllant, *M.N.*, **106**, 130, 1946.

The present process is regarded as explaining the corpuscular emission responsible for the 'great' magnetic storms and for the majority of strong auroral displays. The mean time of transit of these particles between the sun and the Earth is $\sim 1\cdot 5$ days.

(iii) *The ejection of particles from regions of intense coronal radiation*

It is probable that regions of intense coronal emission correspond to regions of abnormally high q (see section 28). Such localized regions arise through the magnetic focusing of the incoming accreted material. According to (16·11) the temperature in region 2 is $\sim V^2/6\mathfrak{R}$ when $q \gg 1$. The mean thermal energy $m_p V^2/2$ of a hydrogen atom (proton + electron) at this temperature is small compared with the energy $m_p V^2(1 + q)/2$ required for escape *along a radial direction.* Thus the proportion of hydrogen atoms with sufficient energy to escape along radial directions is $\sim e^{-(1+q)}$, which is $\ll 1$ when $q \gg 1$. It follows therefore that, even when q is large, there is little tendency for the material in region 2 to evaporate *radially* outwards. A different situation occurs, however, for atoms that move at appreciable angles with the radial direction. Such atoms may leave the localized region of high q and enter regions where the energy required for escape is only $\sim m_p V^2/2$. An appreciable fraction of these atoms are able to escape from the sun. Thus we expect the localized regions of high q to produce a 'splash' in which the material of region 2 is sprayed out from the area in question into regions of small q. This sideways evaporation agrees with the recent observations of Shapley and Roberts, which were discussed in section 9 (iv) in connexion with the properties of M-regions.

Particles emitted in accordance with the present process reach the Earth with velocities $\sim V$. Thus the mean transit time between the sun and the Earth ~ 3 days.

It can be shown that, since the density of accreted atoms in regions of low q is small compared with the density of the solar atmosphere at height h^*, the requirement discussed in (i) is satisfied in the present process.

(iv) *The magnetic condition*

It has been assumed throughout the above discussion that the condition (5·11) is satisfied. As an example we apply this condition to the process discussed in (ii). Thus, putting $\rho = 3 \times 10^{-8}/c$, we obtain the requirement that $|H|$ must be $< \sim 100$ gauss in regions that are ejecting calcium atoms. A similar restriction applies to the ejection of particles from regions of high q.

27. Magnetic Storms and Auroræ†

If the sun were *in vacuo* the magnitude of the general magnetic field would vary approximately as r^{-3} $(r > 1)$. Thus, taking $|H_g| \sim 50$ gauss at the photosphere, the field near the Earth would be $\sim 5 \times 10^{-6}$ gauss. A different situation occurs when ionized accreted material is flowing into the sun. The work of section 5 (iv) shows that in this case the magnetic lines of force must move with the ionized material. Thus the effect of accretion is to push the magnetic lines of force back towards the sun. Important magnetic effects occur, however, when a beam of corpuscles is emitted from the sun. The accreted material is pushed back and the beam carries magnetic energy to great distances from the sun.

We now approximate by assuming that the beam contains a given number of particles distributed with uniform density within a definite solid angle and extending out to a prescribed distance from the sun (even if the particles are all emitted at effectively the same time there will still in general be a considerable spread along the radial direction owing to velocity variations from one particle to another). It can then be shown that the magnetic field within the beam depends on the distance r from the sun according to the factor $\sim ar^{-3/2}$, where a is a coefficient $\sim 0{\cdot}1$. This factor expresses the condition that, as the beam spreads, the magnetic energy per particle must remain approximately constant. Thus, if we take the field at $r = 1$ to be ~ 50 gauss, the field at distance r is $\sim 5/r^{3/2}$ gauss. Putting $r \sim 215$ we obtain a field $\sim 10^{-3}$ gauss near the Earth. Unfortunately this magnetic field cannot be directly measured by an observer on the Earth, since the storm field in the immediate vicinity of the Earth is dominated by electric currents induced in the ionized beam by the magnetic field of the Earth (S. Chapman and J. Bartels, *op. cit.*). Nevertheless, the intrinsic magnetic energy carried by the beam is important at appreciable distances from the

† S. Chapman and J. Bartels, *Geomagnetism*, Vols. I and II, Oxford, 1940.

Earth, where it provides a plausible mechanism for the origin of auroræ.

Let $\boldsymbol{H}_E$, $\boldsymbol{H}_s$ refer respectively to the magnetic field of the Earth and to the field carried from the sun by the corpuscular stream. Since the lines of force of the field $\boldsymbol{H}_E$ move relative to the lines of force of the field $\boldsymbol{H}_s$ we may apply the considerations of section 23. In particular the electric currents flowing in the material within the beam must be strongly concentrated near the neutral points of the field $(\boldsymbol{H}_E + \boldsymbol{H}_s)$. Taking $\boldsymbol{H}_E$ as the field due to a dipole and $\boldsymbol{H}_s$ as a constant field in the neighbourhood of the Earth, it can be shown that for $|\boldsymbol{H}_s| \sim 10^{-3}$ gauss, there are two such neutral points at equal distances $\sim 4 \times 10^9$ cm. from the Earth and lying in a plane parallel to $\boldsymbol{H}_s$ through the magnetic axis of the Earth. The line joining these points passes through the centre of the Earth.

The electric field measured by an observer moving with material near a neutral point is in general $\sim |\boldsymbol{u}| . |\boldsymbol{H}_s|/c$, where $\boldsymbol{u}$ is the velocity of the material relative to the sun. The value of $|\boldsymbol{u}|$ for the corpuscles is $\sim 10^8$ cm./sec. Thus, putting $|\boldsymbol{H}_s| = 10^{-3}$ gauss, the field is $\sim 10^{-3}$ volts/cm. The dimensions of the accelerating region in which this field operates may be taken as having an order of magnitude of 1% of the distance of the neutral point from the centre of the Earth. Thus the accelerating region has dimensions $\sim 4 \times 10^7$ cm., which is small compared with the mean free paths of the particles within the beam. It follows, therefore, that electrons and protons passing near a neutral point are accelerated through a distance 4×10^7 cm. in a field $\sim 10^{-3}$ volts/cm. After leaving the accelerating region the particles move along magnetic lines of force, some of which lead into the Earth's atmosphere. Accordingly, we expect particles with energies $\sim$40,000 eV. to enter the Earth's atmosphere. These energies are of the order necessary to give penetration by electrons to heights $\sim$100 km. above the surface of the Earth.

When $\boldsymbol{H}_s$ is given the positions of the neutral points are determined, since $\boldsymbol{H}_E$ is a known vector ($\boldsymbol{H}_E$ varies with time). The two lines of magnetic force that start from the Earth and pass through the neutral points determine four definite points on the Earth's surface where an auroral display will occur. Two of these points are in the northern hemisphere and the other two are corresponding points in the southern hemisphere. Since $\boldsymbol{H}_s$ varies both in magnitude and direction, the places where the auroral displays occur must vary from one magnetic storm to another.

28. The Ionosphere †

(i) *Observational data*

In the present section we shall be concerned with the ionization of the Earth's atmosphere arising from radiation emitted by the sun. It may be noted, however, that other forms of ionization are also present. In particular, ionization arising from meteorites has been discussed by Appleton in connexion with sporadic ionization in the E-layer and also as being responsible for maintaining ionization in the lower part of the F-layer during the night.

The following data ‡ may be taken as giving average estimates for the E-, F_1-, and F_2-layers.

Table 27.

	E-layer	
Altitude	= 120 km.	
Maximum electron density	= $1{\cdot}5 \times 10^5$ per cm.^{3} (day), 10^4 per cm.^{3} (night).	
Apparent recombination coefficient	= 10^{-8} cm.^{3} sec.$^{-1}$ (day and night).	
Scale height	= 10 km.	
Gas density	= 6×10^{12} particles per cm.^{3}	
Temperature	= 500 deg. K.	
	F_1-layer	
Altitude	= 220 km.	Merged with the F_2-layer at night.
Maximum electron density	= $2{\cdot}5 \times 10^5$ per cm.^{3}	
Apparent recombination coefficient	= 4×10^{-9} cm.^{3} sec.$^{-1}$	
Scale height	= 30 km.	
Gas density	= 10^{11} particles per cm.^{3}	
Temperature	= 1000 deg. K.	
	F_2-layer	
Altitude	= 300 km.	
Maximum electron density	= 10^6 per cm.^{3} (day), $2{\cdot}5 \times 10^5$ per cm.^{3} (night).	
Apparent recombination coefficient	= 8×10^{-11} cm.^{3} sec.$^{-1}$ (day), 3×10^{-10} cm.^{3} sec.$^{-1}$ (night).	
Scale height	= 70 km.	
Gas density	= 2×10^{10} particles per cm.^{3}	
Temperature	= 2000 deg. K.	

† For a thorough discussion see D. R. Bates and H. S. W. Massey, *Proc. Roy. Soc.*, **187**, 261, 1946.

‡ D. R. Bates and H. S. W. Massey, *loc. cit.*

The electron densities, which are maximum values with respect to the diurnal variation, change both throughout the year and with the sunspot cycle. Thus Appleton and Naismith find that there is an increase by factors of 1·50 and 1·56 for the *E*- and F_1-layers respectively between spot minimum and spot maximum, while Appleton gives an increase by a factor of 4·0 for the F_2-layer.† The main yearly changes occurring in the *E*- and F_1-layers are in accordance with the normal effect arising from the inclination of the Earth's axis. The variations in the F_2-layer, on the other hand, are of a far more complex nature.

The apparent recombination coefficient α is defined by the equation

$$\frac{dn_e}{dt} = I - \alpha n_e^2, \tag{28·1}$$

where *I* is the ionization rate per cm.³ This definition of α is independent of the intricate recombination processes that take place in the ionosphere. In the *E*- and F_1-layers α may be taken as independent of time, but not in the F_2-layer. To a sufficient approximation we may put $I = \alpha n_e^2$ during the daytime in the *E*- and F_1-layers. Thus, since α, n_e are known from observation, the value of *I* can be deduced. It is seen that in these layers *I* must vary by a factor $\sim$2·3 over the sunspot cycle. The corresponding variation for the F_2-layer cannot be obtained in a similar manner owing to uncertainty concerning the variation of α.

(ii) *Requirements arising from the observational data*

The following table gives the ionization potentials of the first few levels of the main atmospheric gases.

Table 28.

Gas	O_2	O	N_2	O_2	O_2	O	O_2	O	N_2
Ionization level	1	1	1	2	3	2	4	3	2
Energy required for ionization (eV.)	12·2	13·5	15·5	16·1	16·9	16·9	18·2	18·5	18·7

The corresponding cross-sections are shown in Table 29.

† Appleton and Naismith, *Phil. Mag.*, **27**, 144, 1939. Appleton, *Occ. Notes, R.A.S.*, **3**, 33, 1939.

Table 29.

Gas	Energy of ionizing quanta (eV.)	Continuous absorption cross-section
O	13·5–16·9	$4·5 \times 10^{-18}$
O	16·9–18·5	$1·1 \times 10^{-17}$
O	18·5–25	$1·6 \times 10^{-16}$
O_2	12·2	$\sim 10^{-20}$
O_2	16·1–25	10^{-16}–10^{-17}
N_2	15·5–18·7	10^{-17}
N_2	18·7–25	10^{-16}

These values may be compared with the following table, which gives the products of gas density and scale height (as given in Table 27) for the E-, F_1-, and F_2-layers.

Table 30.

Layer	E	F_1	F_2
Product of gas density and scale height	6×10^{18}	3×10^{17}	$1·4 \times 10^{17}$

From this comparison we conclude:

(*a*) That no ionization process given in Table 28 is capable of explaining the formation of the E-layer. This conclusion is strengthened by work on the composition of the atmosphere,† which shows that, at the height of the E-layer, oxygen is largely in atomic form. Thus the ionization of O_2 at the first level seems to be ruled out on photochemical grounds alone.

(*b*) That the F_1- and F_2-layers are formed by ionization processes with cross-sections $\sim 10^{-17}$. The rate of production of electrons must be approximately the same in both layers. Thus the observed difference of the electron densities must be due to the difference of the recombination coefficient (see Table 27). In this connexion it may be noted that the anomalous behaviour of the F_2-layer is probably due to variations in the recombination coefficient rather than to a variation in the rate of production of electrons. Although it is premature, until the recombination processes are better understood, to speculate on the origin of

† O. R. Wulf and L. S. Deming, *Terr. Mag. and Atmos. Elect.*, **43**, 283, 1938. R. C. Majumdar, *Ind. J. Phys.* **21**, 75, 1938.

these variations, it may nevertheless be of interest in this connexion to consider the effect of neutral hydrogen atoms and molecules entering the Earth's atmosphere from without. This process, which leads to temperature changes in the F_2-layer, may introduce complex changes during the year, since material accreted by the sun is unlikely to be distributed with strict isotropy at the Earth's distance.

Information has recently been obtained † concerning the origin of the *E*-layer by working out the continuous absorption cross-sections for quanta with energies > 20 eV. It has been found that the only satisfactory photon group has an energy spectrum ranging from ~300 eV. up to the nitrogen ionization potential at ~390 eV. The absorption of these quanta leads to the ejection from oxygen atoms of high-energy primary electrons, and the main ionization then arises in collisions between these electrons and the atmospheric gases. The absorption cross-section decreases with increasing energy until the nitrogen ionization potential is reached, when there is a large discontinuous increase in the cross-section. Thus the greatest penetration occurs for quanta with energies just below the nitrogen ionization potential (quanta with energies above the nitrogen ionization potential being absorbed at much greater heights). An important consequence of the discontinuity in the absorption cross-section is to give a sharp base to the *E*-layer. To explain the *E*-layer maximum the intensity of quanta in the range 300 eV. to 390 eV. must increase with increasing energy.

Finally, we note that the energy of emission from the sun of the radiation producing a particular layer can be inferred from the relation $I \sim \alpha n_e^2$ taken together with the scale height of the layer in question. Accepting the above conclusions concerning the quanta responsible for the ionization of the *E*-, F_1-, and F_2-layers, we obtain the requirements shown in Table 31.

Table 31.

Energy range (eV.)	~300 — ~390	13·5 — ~20
Required emission from the surface of the sun (ergs. per cm.^{2} per sec.)	~10^3	~5×10^3

† D. R. Bates and F. Hoyle, *Terr. Mag. and Atmos. Elect.*, **53**, 51, 1948.

(iii) *Comparison with the solar data*

(*a*) *The E-layer*

We observe at once that the ionization potentials of heavy elements present in the corona, as given in Table 7, are of the required order of magnitude. Furthermore, (17·14), (17·15) give, for $x < \sim 500$ eV., the form of the emission spectrum due to iron. Taking iron to be the most abundant heavy element in the corona, we require that the integrals with respect to x of these distributions shall give a mean intensity $\sim 10^3$ ergs per cm.^{2} per sec. This condition leads to the result that $(c_{Fe} + \boldsymbol{c}_{Fe})$ must be $\sim 10^{-3}$ in the corona, which exceeds the concentration of iron at the photosphere by a factor ~ 10. This implies that the accreted material would have to be richer in heavy elements than the solar material (previous discussions, in which the accreted material was regarded as composed essentially of hydrogen, are not affected by heavy element concentrations $\sim 10^{-3}$). The present theory stands or falls according as to whether or not this requirement is satisfied.

If the heavy elements in the corona are mainly supplied by the accreted material c_{Fe}, $\boldsymbol{c}_{Fe}$ must increase with height. This variation affects the form of (17·14), (17·15), since the dependence of c_{Fe}, $\boldsymbol{c}_{Fe}$ on height can be represented as a dependence on x. For a sufficiently rapid increase of c_{Fe}, $\boldsymbol{c}_{Fe}$ with x the intensity of the photon group also increases with $x (x < \sim 500$ eV.). Beyond ~ 500 eV. the intensity falls rapidly on account of the x^{-5} factor, discussed in section 17. Accordingly, the present work agrees with the requirements discussed above.

The suggestion that the accreted material is richer in heavy elements than normal solar material has an interesting application to Waldemeier's correlation between *M*-regions and regions of intense coronal emission (see section 9 (iv)). It was shown in section 26 that corpuscular emission from *M*-regions can be explained in terms of localized regions of high q, produced by the focusing of accreted material. We now see that such regions will show exceptionally strong emission if the accreted material is rich in heavy elements. There would be no such correlation if the accreted material were poorer in heavy elements than the normal solar material.

(*b*) *The F_1- and F_2-layers*

The requirements set out in (ii) are in good agreement with the results of section 17, as given in Fig. 6.

(*c*) *Variations with the sunspot cycle*

The emission of ionizing radiation, as given in section 17, is proportional to q^2. We have seen that one of the effects of local magnetic fields in the surface layers of the sun is to disturb the isotropy of the accretion process. Thus at some places accreted material is focused so as to give a localized region of high q. While such disturbances do not alter the average value of q taken over the whole solar surface, the average value of q^2, on the other hand, increases as the deviations from isotropy increase. It follows therefore that the magnetic focusing of accreted material provides a process that gives a variable average emission intensity. Since magnetic effects are strongest at sunspot maximum, the emission increases between spot minimum and spot maximum.

CHAPTER VII

THE EMISSION OF RADIO WAVES FROM THE SUN

29. The Emission under Thermodynamic Conditions

(i) *In the absence of a magnetic field*

Under thermodynamic conditions a body at temperature θ emits radio waves in the frequency range ν to $(\nu + d\nu)$ according to the well-known expression

(29·1) $2\pi k\theta\nu^2 d\nu/c^2$ ergs per unit surface area per sec.

On account of the variations of temperature in the solar atmosphere it is not immediately clear what values of θ must be used in applying (29·1) to the case of the sun. The following considerations show that the appropriate value of θ varies with ν.

For the present purpose it is convenient to average over the ascending and descending columns in the corona. Thus, now regarding T, n_e as such average values, we take

$$(29\cdot2)\qquad \left.\begin{aligned} T &\sim 2 \times 10^6 h/h^* \text{ deg. K.,} \\ n_e &\sim 10^8 (h^*/h)^{3/2} \text{ electrons/cm.}^3, \end{aligned}\right\} h < h^*$$

in accordance with the results of section 16.

The absorption coefficient for quanta of frequency ν in ionized hydrogen is †

$$(29\cdot3)\qquad \frac{8\pi\epsilon^6 n_e^2}{3\sqrt{3}\hbar c(2\pi m_e)^{3/2}(kT)^{\frac{1}{2}}\nu^3},$$

where ϵ, m_e are the electronic charge and mass respectively, and $\hbar$ is Planck's constant divided by 2π. It is well known that this expression must be modified, so as to take proper account of stimulated emission, in order to obtain the opacity. The necessary modification consists in multiplying (19·3) by

$$\{1 - exp(-2\pi\hbar\nu/kT)\},$$

† J. A. Gaunt, *Phil. Trans. Roy. Soc.*, **229**A, 163, 1930.

which can be written to sufficient accuracy as $2\pi\hbar\nu/kT$ for radio wave frequencies. Accordingly, the opacity, k_ν say, is given by

$$k_\nu = \frac{16\pi^2\epsilon^6 n_e^2}{3\sqrt{3}c(2\pi m_e kT)^{3/2}\nu^2}. \tag{29·4}$$

The physical significance of the height h^+ defined by

$$\int_{h^+}^{\infty} k_\nu(h)dh = 1 \tag{29·5}$$

is that almost all quanta of frequency ν escaping from the sun are emitted by material situated at heights greater than this particular value of h. Quanta emitted at appreciably lower heights do not escape (except in rare cases) on account of absorption in overlying material. For $\nu > 5 \times 10^7$ cycles/sec. it is sufficiently accurate to use (29·2) for n_e even when $h > h^*$. The value of h^+ given by (29·5) is then easily worked out by using (29·2) and (29·4). The required result is

$$h^+/h^* = 1 \cdot 3 \times 10^4 \nu^{-4/7}. \tag{29·6}$$

The following table gives a selection of numerical values for the cases that are of interest.

Table 32.

ν (cycles per sec.)	5×10^7	10^8	2×10^8	5×10^8	3×10^9	10^{10}	3×10^{10}
h^+ (km.)	66,000	44,000	30,000	18,000	5000	3500	2000
$T(h^+)$ in deg. K.	$1 \cdot 0 \times 10^6$	$6 \cdot 8 \times 10^5$	$4 \cdot 6 \times 10^5$	$2 \cdot 8 \times 10^5$	10^5	$4 \cdot 8 \times 10^4$	$2 \cdot 6 \times 10^4$
$\int_{h^+}^{\infty} n_e(h)dh$	$3 \cdot 7 \times 10^{18}$	$4 \cdot 5 \times 10^{18}$	$5 \cdot 4 \times 10^{18}$	$7 \cdot 1 \times 10^{18}$	$1 \cdot 2 \times 10^{19}$	$1 \cdot 7 \times 10^{19}$	$2 \cdot 3 \times 10^{19}$

The rate of emission of radio waves at these frequencies is given, in the absence of a magnetic field, by inserting the appropriate values of $T(h^+)$ in place of θ in (19·1).

(ii) *In the presence of a magnetic field*

The component parallel to the magnetic field $\boldsymbol{H}$, of the velocity $\boldsymbol{u}$ of an electron, is unaffected by the magnetic field. On the other hand, the projection of the orbit of an electron on a plane perpendicular to $\boldsymbol{H}$ may to a sufficient approximation be taken as a circle of radius

$$m_e c\,|\,\boldsymbol{u} \times \boldsymbol{H}\,|/\epsilon H^2. \tag{29·7}$$

It can be verified that in all cases of interest this expression is of the order of, or less than, 1 cm. Over such distances $\boldsymbol{H}$ can evidently be regarded as a constant vector. The circular motion is periodic with frequency ν_0 given by

$$\nu_0 = \epsilon \,|\, \boldsymbol{H} \,|/2\pi m_e c. \tag{29·8}$$

It is seen that the orbital period is independent of the velocity of the electron and depends only on the magnitude of the magnetic field. The following table shows the relation between ν_0 and $|\,\boldsymbol{H}\,|$.

Table 33.

ν_0 (cycles per sec.)	5×10^7	10^8	2×10^8	5×10^8	3×10^9	10^{10}	3×10^{10}
$\|\,\boldsymbol{H}\,\|$ (gauss)	19	38	77	192	1153	3847	11,536

Since (29·7) is very small compared with the mean free paths of electrons in the solar atmosphere, it follows that an electron completes a large number of oscillations between successive collisions with other particles. Thus the electrons can to good approximation be regarded as radiating at the rate given by an electron moving continuously in a circular orbit. This is given by

$$2(2\pi\nu_0\epsilon \,|\, \boldsymbol{u} \times \boldsymbol{H} \,|)^2/3c^3\boldsymbol{H}^2, \tag{29·9}$$

which is twice the emission rate for a charge ϵ oscillating with maximum speed $|\,\boldsymbol{u} \times \boldsymbol{H}\,|/|\,\boldsymbol{H}\,|$ in simple harmonic motion of frequency ν_0.† The energy of emission in the frequency range ν to $(\nu + d\nu)$ is given ‡ by multiplying (29·9) by the factor

$$\frac{4\gamma d\nu}{16\pi^2(\nu - \nu_0)^2 + \gamma^2}, \tag{29·10}$$

where $\gamma = 8(\pi\epsilon\nu_0)^2/3m_e c^3$. The ratio γ/ν_0 is $7{\cdot}5 \times 10^{-12}$ for $\nu_0 = 3 \times 10^{10}$ cycles/sec., and is correspondingly smaller at lower frequencies. Thus the line emitted by a single electron is extremely sharp. On the other hand, the combined emission by a large number of electrons, all oscillating with frequency ν_0, is much wider on account of thermal broadening. The emission along the direction of $\boldsymbol{H}$ is circularly polarized, is linearly polarized

† M. Abraham and R. Becker, *Classical Theory of Electricity and Magnetism*, Blackie, 1932, p. 223. The motion in a circular orbit is equivalent to two perpendicular linear oscillations of equal amplitude in quadrature.

‡ W. Heitler, *The Quantum Theory of Radiation*, Oxford, 1936, p. 34. In the present work ν_0 is smaller by a factor 2π than the ν_0 used by Heitler.

in directions perpendicular to $\boldsymbol{H}$, and is elliptically polarized in intermediate directions.

The rate of absorption by an electron moving with frequency ν_0 in a circular orbit in an unpolarized isotropic radiation field with energy density $I(\nu)$ per unit frequency range is given by †

$$2\pi\epsilon^2 I(\nu_0)/3m_e. \tag{29·11}$$

It is important to notice that absorption involves a polarization effect similar to the case of emission. Thus, if we consider radiation moving in a given direction, the absorbed radiation has the same polarization as the radiation emitted in the direction by the electron in question. The natural half-width of the absorption line of a single electron is again γ, but the absorption line arising from a large number of electrons is much wider on account of thermal broadening.

It can be shown that (29·9), (29·11) represent emission and absorption rates that are very large compared with the rate occurring in the absence of a magnetic field. Accordingly, provided we consider a frequency range corresponding to the variation of magnetic intensity occurring in the corona, thermodynamic equilibrium between radio waves and matter must be attained at electron densities much lower than the values obtained in (i). Taking the magnetic intensity as variable over the range 0 to $\sim$50 gauss, it is seen from Table 33 that this frequency range extends up to $\sim 1{\cdot}3 \times 10^8$ cycles/sec. It follows that for these frequencies the effect of the general solar magnetic field is to produce a marked increase in the opacity k_ν. This increase shifts h^+ from the inner corona to the outer corona. The value of θ to be used in (29·1) is therefore $\sim 2 \times 10^6$ deg. K. The considerations of (i) still apply, however, for frequencies appreciably greater than $1{\cdot}3 \times 10^8$ cycles/sec.

(iii) *Comparison with observational data for a 'quiet' sun*

Experimental results are usually given in terms of an 'effective' temperature θ_{eff}, which is defined by the requirement that an emission

$$2\pi k\theta_{eff}\nu^2 d\nu/c^2 \text{ per unit surface area per unit time,} \tag{29·12}$$

† This result can be obtained by a method similar to that given in W. Heitler, *op. cit.*, p. 39.

occurring uniformly over the whole solar disk, must give the observed intensity. (It is assumed that there is no absorption between the sun and the earth.) The work of (i) and (ii) shows that, for $\nu \sim 3 \times 10^{10}$ cycles/sec., θ_{eff} must be of the order of the temperature near the base of the chromosphere. As ν decreases, θ_{eff} increases until for $\nu \sim 1{\cdot}3 \times 10^{8}$ cycles/sec. it is $\sim 2 \times 10^{6}$ deg. K. The effective temperature then remains approximately constant as ν decreases below $1{\cdot}3 \times 10^{8}$ cycles/sec. These results are in good agreement with the observational data for a 'quiet' sun.†

30. Emission under Disturbed Conditions

(i) *Experimental data*

Disturbed conditions are correlated with solar activity and probably in particular with solar flares. During these conditions it is found that θ_{eff} attains a mean level $\sim 10^{9}$ deg. K. in the metre wave-band (no corresponding enhancement being observed in the centimetre wave-band). When a particular wavelength in this band is examined it is found that rapid fluctuations in θ_{eff} occur. These fluctuations become more marked as the wavelength in question increases up to ~ 5 metres, but beyond 5 metres there is an apparent decrease which is probably due to absorption occurring in the Earth's ionosphere. The highest values of θ_{eff} which are $\sim 10^{12}$ deg. K., occur in short bursts that usually last for only a few seconds. It is probable that there is no detailed correlation between bursts on different frequencies.‡

The departure from the thermodynamic rate of radiation is even more marked than is suggested by these results. It will be recalled that the definition of θ_{eff} is based on uniform emission over the whole solar surface, whereas Ryle and Vonberg § have shown that at $\nu = 1{\cdot}75 \times 10^{8}$ cycles/sec. the enhanced radiation originates in an area considerably smaller than the solar disk. The radiation observed by these authors was circularly polarized. A similar result has also been found by Martyn at $\nu = 2 \times 10^{8}$ cycles/sec., and by Appleton and Hey at $\nu = 6{\cdot}4 \times 10^{7}$ cycles/

† Reber, *Ap.J.*, **100**, 279, 1944. Southworth, *J. Frank. Inst.*, **239**, 285, 1945, and **241**, 167, 1946. Dicke and Beringer, *Ap.J.*, **103**, 375, 1946. Pawsey, Payne-Scott, and McCready, *Nature*, **157**, 158, 1946. D. F. Martyn, *Nature*, **158**, 632, 1946.

‡ Williams and Hands, *Nature*, **158**, 511, 1946. Lovell and Banwell, *Nature*, **158**, 518, 1946.

§ M. Ryle and D. D. Vonberg, *Nature*, **158**, 339, 1946.

sec.† Although enhanced radiation is probably always polarized, circular polarization represents a special case.

(ii) *Theoretical discussion*

The occurrence of emission rates far above the thermodynamic level raises an important theoretical problem. Kiepenhauer ‡ has argued that the enhanced radiation arises from the large radiation rate given by (29·9). This suggestion must be abandoned, however, when, as Kiepenhauer supposes, the velocities of the electrons are of thermal origin. For, so long as the material is at a definite kinetic temperature, thermodynamic results apply quite independently of the processes leading to equilibrium between matter and radiation. This general principle of statistical mechanics will be verified for the present case.

Consider n_e electrons/cm.³ emitting radiation of frequency ν_0. Then the total emission due to the orbital motion about the magnetic lines of force is

$$(30\cdot1)\qquad 2n_e(2\pi\nu_0\epsilon)^2\,\overline{|\boldsymbol{u}\times\boldsymbol{H}|^2}/3c^3H^2,$$

where $\overline{|\boldsymbol{u}\times\boldsymbol{H}|^2}/H^2$ is the mean square of the component of velocity perpendicular to $\boldsymbol{H}$. The rate of absorption from an isotropic unpolarized radiation field of energy density $I(\nu)$ per unit frequency range is

$$(30\cdot2)\qquad 2\pi n_e\epsilon^2 I(\nu_0)/3m_e.$$

If n_e is sufficiently large $I(\nu_0)$ must reach an equilibrium value such that (30·1) and (30·2) become equal. This condition gives

$$(30\cdot3)\qquad I(\nu_0)=4\pi\nu_0^2 m_e\,\overline{|\boldsymbol{u}\times\boldsymbol{H}|^2}/c^3H^2,$$

and if the electron velocities are of thermal origin (30·3) reduces to

$$(30\cdot4)\qquad I(\nu_0)=8\pi\nu_0^2kT/c^3,$$

since $m_e\overline{|\boldsymbol{u}\times\boldsymbol{H}|^2}/H^2$ is then equal to $2kT$. This result agrees with the thermodynamic energy density per unit frequency range for the case $2\pi\hbar\nu_0 \ll kT$. If n_e is not sufficiently high for equilibrium to be attained then $I(\nu_0)$ is less than the value given by (30·4). It can be shown, by considering the thermal broadening of the emission and absorption lines, that these considerations are

† D. F. Martyn, *Nature*, **158**, 308, 1946. E. V. Appleton and J. S. Hey, *Nature*, **158**, 339, 1946.

‡ K. O. Kiepenhauer, *Nature*, **158**, 340, 1946.

not affected by the variations of intensity occurring in the solar magnetic field.

The thermodynamic criteria can in principle be overcome if the radiation field is not in random phase relative to each oscillating electron. It is difficult, however, to devise a process in which the necessary phase relations are satisfied. Recently, Shklovsky † has suggested that a co-phased electron system oscillating under Coulomb forces might occur in the corona. The frequency of such oscillations is given by the Langmuir formula

$$(4\pi\epsilon^2 n_e/m_e)^{\frac{1}{2}}. \tag{30·5}$$

But, although the values of n_e occurring in the corona lead to frequencies in the metre wave-band, the required phase relations are not satisfied because the electron system is large compared with the electromagnetic wavelength. Moreover, the emission of circularly polarized waves presents a difficulty in such a theory.

An interesting possibility ‡ arises in connexion with the runaway electrons discussed in section 24. These electrons acquire energies $\sim$10,000 eV. in the neighbourhood of a neutral point of the magnetic field. After leaving the accelerating region they form a group not in thermal equilibrium with surrounding material. The energy density per unit frequency range of radio-waves emitted by such a group is given by putting

$$m_e \overline{|\boldsymbol{u} \times \boldsymbol{H}|^2}/H^2 = 10{,}000 \text{ eV.} \tag{30·6}$$

in (30·3). This leads to $\theta_{eff} \sim 10^8$ deg. K. for radio-wave frequencies emitted by the runaway electrons. These frequencies are determined by the range of magnetic intensity occurring in the region through which the electrons are moving. For example, if the magnetic intensity is $\sim$30 gauss throughout the region in question, then, according to (29·8), the enhanced radiation occurs only for frequencies close to $\nu = 8 \times 10^7$ cycles/sec. Excluding the special case in which the magnetic lines of force are approximately perpendicular to the line of sight, it follows from the large mean value of $\boldsymbol{u} \,.\, \boldsymbol{H}/|\boldsymbol{H}|$ that the line emitted by the runaway electrons is much broader than the absorption line of surrounding material. Hence only a small fraction of the enhanced radiation is absorbed in the surrounding material. Furthermore,

† J. S. Shklovsky, *Nature*, **159**, 752, 1947.
‡ R. G. Giovanelli, *Nature*, **161**, 133, 1948.

it can be shown that provided the magnetic intensity changes monotonically with height above a given level, then not more than about 50% of the enhanced radiation emitted at this level can be absorbed by material at appreciably greater heights.

It is important to notice that the value 10,000 eV., obtained in section 24, is by no means an upper limit for the energy of a runaway electron. This particular estimate was based on an electric field $\sim 10^{-3}$ volts/cm. near a neutral point, and this value was obtained by taking $\sim 3 \times 10^3$ cm./sec. for the relative velocity between the general solar magnetic field $\boldsymbol{H}_g$ and a spot field $\boldsymbol{H}_s$. Such a relative velocity may be appreciably exceeded if the neutral point rises appreciably above the reversing layer. For, velocities comparable with the speed of convection currents in the lower chromosphere may then be expected to occur. Thus a relative velocity $\sim 10^6$ cm./sec. would provide runaway electrons with energies $\sim 3 \times 10^6$ eV. The corresponding value of $\theta_{eff} \sim 3 \times 10^{10}$ deg. K., which is adequate to explain all but the most intense of the observed bursts.

APPENDIX I

NOMENCLATURE AND LIST OF SYMBOLS

'ln' denotes logarithm to the base *e*.

'log' denotes logarithm to the base 10.

Unless otherwise stated all numerical calculations are in terms of c.g.s. units.

The symbol $\sim$ means 'about' or 'of the order of'.

M is the solar mass and is $1{\cdot}985 \times 10^{33}$ gr.

R is the radius of the photosphere and is $6{\cdot}951 \times 10^{10}$ cm.

The mean distance of the earth from the sun is $1{\cdot}49 \times 10^{13}$ cm.

$\boldsymbol{H}$ denotes a magnetic field.

$\boldsymbol{H}_g$ is the general magnetic field of the sun.

$\boldsymbol{H}_s$ is the magnetic field of a sunspot (and also the geomagnetic storm field in section 27).

$\boldsymbol{H}_E$ is the magnetic field of the earth.

$\boldsymbol{E}$ denotes an electric field (E_m is used in section 24 for the component parallel to the magnetic field).

$\boldsymbol{A}$ is a vector potential.

σ is the conductivity when the electric and magnetic fields are parallel (and is also used for collision cross-sections).

When the electric and magnetic fields are perpendicular σ^I is the conductivity in the direction of the electric field and σ^{II} is the conductivity in a direction perpendicular to both the electric and magnetic fields.

k is Boltzmann's constant, $1{\cdot}372 \times 10^{-16}$ (and is also a quantity $\sim 6{\cdot}8 \times 10^{13}$ in (5·5) and (5·5′)).

$\boldsymbol{j}$ is the electric current density.

$\boldsymbol{u}$, $\boldsymbol{v}$ are velocity vectors.

G is the constant of gravitation, $6{\cdot}66 \times 10^{-8}$.

c is the velocity of light, $3{\cdot}00 \times 10^{10}$ cm. per sec.

P is the total hydrostatic pressure.

P_e is the electron pressure.

r is the distance from the solar centre.

h is a height co-ordinate measured from a certain level in the reversing layer (and is also Planck's constant, $6{\cdot}55 \times 10^{-27}$, in (7·7) for example, and in section 10 represents the angular momentum per unit mass of the accreted material).

$\hbar$ is Planck's constant divided by 2π.

h^* is the height of the base of region 2.

g is solar gravity and is $2{\cdot}736 \times 10^4$ at the photosphere.

V is the parabolic velocity at height h^* and is $\{2GM/(R+h^*)\}^{\frac{1}{2}}$.

$4\pi R^2 \tau V$ is the rate at which accreted hydrogen atoms enter the solar atmosphere.

τ_∞ is the space density of interstellar hydrogen atoms at a large distance from the sun.

v is the velocity of the sun relative to the interstellar material (and also in 7·1 for an electron velocity).

m_p, m_e are the masses $1{\cdot}66 \times 10^{-24}$, $9{\cdot}01 \times 10^{-28}$ of the proton and electron.

n is the space density of hydrogen atoms.

n_p is the space density of protons.

n_e is the space density of electrons.

n_Z is the space density of positive ions of charge $-Z\epsilon$.

ϵ is the charge of the electron $-4{\cdot}77 \times 10^{-10}$.

x is an ionization potential (sometimes of hydrogen).

s is the principle quantum number of hydrogen.

$\mathfrak{R}$ is the gas constant, $8{\cdot}26 \times 10^7$.

Θ is an average temperature in region 2.

$N(h^*)$ is the number of hydrogen atoms lying above height h^* in a radial column of cross-section 1 cm.^{2}

q is a numerical factor measuring the ratio of the downward acceleration on hydrogen atoms in region 2 due to accretion and that due to solar gravity, and is equal to

$$0{\cdot}625\, G^2M^2V\tau_\infty/gv^3R^2N(h^*).$$

$T(h)$ is the temperature at height h in a descending column in region 1.

$\rho(h)$ is the density at height h in a descending column in region 1.

$w(h)$ is the velocity of material in a descending column in region 1.

$S(h)$ is the average cross-section of a descending column.

$U(h)$ is the rate of transfer at height h between the downward columns and the ascending columns.

$\boldsymbol{T}$, $\boldsymbol{\rho}$, $\boldsymbol{w}$, $\boldsymbol{S}$, $\boldsymbol{U}$ are the corresponding quantities for an ascending column (these quantities are *not* vectors).

T^* is the temperature at height h^* in a descending column.

α, β, $\boldsymbol{\alpha}$, $\boldsymbol{\beta}$ are dimensionless parameters.

K is a constant with dimensions mass/(time)^3 (and is also used to denote the absolute temperature scale).

H is a theoretical scale height used in (11·9) and defined by $H = \mathfrak{R}\Theta/g\mu(1+q)$.

H is used as an empirical scale-height in section 6.

μ is a mean molecular weight (and is also used for GM in section 10).

λ is a wavelength (and is also used for a dimensionless parameter).

ν is a frequency.

γ is a natural half-line width.

Q is used to denote the rate of transfer of material in region 1 (and is also used for collision cross-sections).

η is an electron energy.

APPENDIX II

THE ORIGIN OF THE MAGNETIC FIELD OF THE SUN

(i) *Preliminary remarks*

There is an important difference in principle between the above work and the problem of the origin of the solar magnetic field. So far it has only been necessary to enquire into present properties, whereas we shall now be concerned with the past history of the sun.

The time required for the magnetic field, produced by any given generating process, to attain a steady value is $\sim\sigma^I R^2/c^2$ (see section 5 (v)). Since the conductivity depends on the distance r from the solar centre, the time of build-up of a steady magnetic field must vary between different parts of the sun. Cowling has given † the values of $\sigma^I(r)/c^2$ shown in Table 34. It follows from these values that the sun is most readily magnetized in the surface layers. But even at the surface a time $\sim 10^6$ years is required for the magnetic field to attain a steady value. This means that the problem of the origin of the solar magnetic field must be considered in the light of the past history of the sun.

Table 34.

r (R as unit)	0	·1	·2	·3	·4	·5	·6	·7	·8	·85	·9	·95	1·0
$10^5\sigma^I/c^2$	76·5	62	36	20·5	12	7·1	3·95	2·05	·9	·5	·3	·1	·001
$\sigma^I R^2/c^2$ (10^9 years as unit)	120	98	57	33	19	11	6·3	3·3	1·4	·79	·48	·16	·0016

It has been suggested that the Quaternary Ice Epoch (occurring on the earth during the last million years) was due to a temporary increase in the kinetic energy $1{\cdot}2 \times 10^6 q$ ergs per cm.^{2} per sec., with which accreted material enters the solar atmosphere.‡ According to meteorological requirements this increase of energy must amount to about 10% of the normal emission from the photo-

† T. G. Cowling, *M.N.*, **105**, 166, 1945.
‡ F. Hoyle and R. A. Lyttleton, *Proc. Camb. Phil. Soc.*, **35**, 405, 1939.

sphere ($\sim 6 \times 10^{10}$ ergs per cm.^{2} per sec.). Thus on this view the value of q must have been $\sim 5 \times 10^3$ during an appreciable fraction of the last million years. That is, the value $q \sim \cdot 25$, deduced in chapter IV, is to be regarded as a recent development. The case $q \sim 5 \times 10^3$ is specially important because it can be shown that the discussion of section 19 (iii), which led to the conclusion that photoelectric ionization prevents neutral atoms reaching the solar atmosphere, becomes invalid on account of electron-proton recombinations occurring in the accreted material. An appreciable proportion of the accreted atoms then enter the solar atmosphere in a neutral form.

(ii) *The generation of an electric current*

It is well known that electric currents can be produced in a gas, even if no external electric field is applied to the system. For example, a current arises in single-stream regions whenever there is a non-zero gradient of the hydrostatic pressure P. It can be shown that in this case the current is equivalent to that produced by an electric field grad $P/\epsilon n_e$.

More complicated effects occur in double-stream regions. In the following work we are concerned with a problem in which ionized hydrogen moves relative to neutral hydrogen, and we take $\boldsymbol{u}$, $\boldsymbol{v}$ as the mean velocities of protons and electrons respectively, relative to the neutral atoms. Furthermore, we assume that the collisions occurring between the ionized and the neutral hydrogen do not set up an electric field, and, for the moment, we neglect the effect of any magnetic field arising from this process. Then the time derivatives of $\boldsymbol{u}$, $\boldsymbol{v}$ are given approximately by

$$(1) \qquad \frac{d\boldsymbol{u}}{dt} \sim -n_n c_{pn} Q_{pn}(c_{pn})\boldsymbol{u} - \frac{m_e}{m_p} n_e c_{ep} Q_{ep}(c_{ep})(\boldsymbol{u} - \boldsymbol{v}),$$

$$(2) \qquad \frac{d\boldsymbol{v}}{dt} \sim -n_n c_{en} Q_{en}(c_{en})\boldsymbol{v} + n_p c_{ep} Q_{ep}(c_{ep})(\boldsymbol{u} - \boldsymbol{v}),$$

where n_n, n_e, n_p are the space densities of neutral atoms, electrons, and protons respectively, c_{pn} is the root mean square velocity of the protons relative to the neutral atoms (the thermal velocities of the protons and neutral atoms affect c_{pn}, but not the mean velocity $\boldsymbol{u}$), c_{ep} is the root mean square velocity of the electrons relative to the protons, c_{en} is the root mean square velocity of the electrons relative to the neutral atoms, $Q_{pn}(c_{pn})$ is the proton-neutral atom collision cross-section corresponding to relative velocity c_{pn}, $Q_{ep}(c_{ep})$ is the electron-proton collision cross-section corresponding to relative velocity c_{ep}, and $Q_{en}(c_{en})$ is the electron-neutral atom collision cross-section corresponding to relative velocity c_{en}.

The application made below of (1), (2) concern the case $c_{en}/c_{pn} \sim (m_p/m_e)^{\frac{1}{2}}$, $c_{en} \gg |\boldsymbol{u}|, |\boldsymbol{v}|$. Then the terms on the right-hand side of (1) are small compared with the terms on the right-hand side of (2). Accordingly, $\boldsymbol{v}$ can change appreciably during a time interval in which $\boldsymbol{u}$ stays effectively constant. We assume, in anticipation of later work, that the ionized hydrogen is exposed to collisions with the neutral atoms for sufficiently long for $\boldsymbol{v}$ to attain a steady value ($d\boldsymbol{v}/dt = 0$), but not long enough for $\boldsymbol{u}$ to change appreciably. Thus $\boldsymbol{v}$ takes the value given by

(3) $$\sim \boldsymbol{u}/\{1 + n_n Q_{en}(c_{en})/n_p Q_{ep}(c_{ep})\}.$$

The current density $\boldsymbol{j}$ is equal to $n_e \epsilon(\boldsymbol{u} - \boldsymbol{v})$, which by (3) becomes (we put $n_e = n_p$)

(4) $$\boldsymbol{j} \sim \frac{\epsilon n_n Q_{en}(c_{en})\boldsymbol{u}}{Q_{ep}(c_{ep})\{1 + n_n Q_{en}(c_{en})/n_p Q_{ep}(c_{ep})\}}.$$

Under the conditions discussed below $n_n/n_p \sim 1$,

$$Q_{en}(c_{en})/Q_{ep}(c_{ep}) \sim 1,$$

so that this expression can be written in the form

(5) $$\boldsymbol{j} \sim \epsilon n_p Q_{en}(c_{en})\boldsymbol{u}/Q_{ep}(c_{ep}).$$

(iii) *The magnetization of the sun*

The discussion of subsections (i) and (ii) indicate that the magnetization of the sun may have occurred during the last million years as a result of a temporary period of very rapid accretion. Although the magnetization process is no longer operative, the values given in Table 34 show that there cannot yet have been a large decay in the intensity of the magnetic field.

First, we note that $\boldsymbol{u}$ cannot be taken as the inward radial velocity $(2GM/r)^{\frac{1}{2}}$ of the accreted atoms, because, as pointed out at the end of section 11, an electric field arises in this case that prevents any flow of current along radial directions. But such an electric field does not arise when there is a suitable horizontal component of relative velocity. Thus, if the accreted material possesses negligible angular momentum about the sun, the effect of solar rotation is to give a relative velocity between the solar atmosphere and the incoming atoms. This relative velocity is directed along lines of solar latitude and is of magnitude $2 \cos \lambda$ km./sec., where λ is the latitude. The resulting electric current does not produce an accumulation of space charge, since the currents are directed along circles of latitude. Accordingly, if u is taken as the relative velocity arising from solar rotation, the flow of current given by (6) does not generate an electric field.

Second, it follows that, since material is convected into region 2 from below, the material of the solar atmosphere in region 2 is continuously supplied with angular momentum about the axis of rotation of the sun. This supply of angular momentum preserves the relative velocity between the material of the solar atmosphere and the incoming accreted atoms. Moreover, since the rate of convection into region 2 from below must be comparable with, or greater than, the rate of accretion, the relative velocity is maintained at a value $\sim 2 \cos \lambda$ km./sec. This justifies the assumption made in (ii) that u may be taken as remaining constant.

In the following work we put n_n equal to the space density of the incoming neutral accreted atoms. Then by (11·8) we have

(6) $$n_n \sim \tau = 8{\cdot}3 \times 10^6 q \text{ atoms/cm.}^3.$$

Moreover, it can be shown from (16·11) that $n_e(r^*)$, $n_p(r^*)$, which represent the densities at the base of region 2 of electrons and protons belonging to the solar atmosphere, are given by

(7) $$n_p(r^*) = n_e(r^*) \sim 3 \times 10^6 q \text{ atoms/cm.}^3.$$

For simplicity, we regard the solar atmosphere in region 2 as being confined to a range of height equal to the scale height H. Then the current generated at latitude λ, by the process discussed in (ii), is

(8) $$\sim 2 \times 10^5 \cos \lambda \, n_p(r^*) H \epsilon Q_{en}(c_{en}) / Q_{ep}(c_{ep}).$$

Now it is well known that the steady magnetic field, produced by such a distribution of current on the surface of a sphere, is uniform within the sphere and has intensity

(9) $$8\pi \times 10^5 n_p(r^*) H \epsilon Q_{en}(c_{en}) / c Q_{ep}(c_{ep}).$$

So far we have neglected the time required for the magnetic field to attain this steady value. During the growth of the magnetic field an induced electric field opposes the current (8). This effect, which was omitted in (ii), reduces the magnetic field below the value given by (9). By a discussion similar to that of section 5 (iii) it can be shown that if the above process operates for a time t ($<\sigma^{I}(r)R^2/c^2$, $0 \leqslant r \leqslant 1$) then the magnetic field builds up to a value

$$(10) \qquad \sim \frac{8\pi \times 10^5 \, . \, ctn_p(r^*)H\epsilon Q_{en}(c_{en})}{\sigma^{I}(r)R^2 Q_{ep}(c_{ep})}, \quad r \leqslant 1,$$

at distance r from the solar centre.

The main contributions to c_{ep} and c_{en} arise from the thermal velocities of the electrons, which are $\sim V(m_p/m_e)^{\frac{1}{2}} = 2 \cdot 4 \times 10^9$ cm./sec. (for the case $q > 1$ see (16·11)). The corresponding thermal energies are ~ 1000 eV., and with such energies

$$Q_{ep}(c_{ep}) \sim Q_{en}(c_{en}).$$

Remembering that $n_p(r^*)H = N(r^*) = 1 \cdot 36 \times 10^{18}$ atoms/cm.^{2}, and taking $t = 5 \times 10^5$ years for the duration of the period of very rapid accretion, we obtain the values given in Table 35.

Table 35.

r (R as unit)	0	·1	·2	·3	·4	·5	·6	·7	·8	·85	·9	·95	1·0
Magnetic intensity (gauss)	·23	·28	·48	·83	1·4	2·5	4·3	8·3	19	34	57	170	17,000

These results are to be regarded as order of magnitude estimates rather than exact values. In particular, we have neglected the ionization occurring in collisions between the neutral atoms and electrons belonging to the solar atmosphere. This process is difficult to investigate quantitatively because a proper analysis involves details of the way in which the incoming accreted atoms mix with the solar atmosphere. As pointed out in section 12 this is, in itself, an intricate problem. It may be noted, however, that since a decrease in the concentration of neutral atoms weakens the magnetic intensity, the electron collisions tend to decrease the magnetization of the sun. This tendency is offset to some extent by our assumption that the solar atmosphere in region 2 is confined to a range of height equal to the scale height H.

(iv) *General considerations*

The above results provide the form of skin magnetization assumed in the work of chapter v. Moreover, the high value of the magnetic intensity at $r = 1{\cdot}0$ suggests that the magnetic fields of sunspots may be generated in the surface layers and subsequently transported by convection to deeper levels (see the remarks of section 22, concerning the lifetimes of the magnetic fields of sunspots). But in this connexion it is noted that convection at and below the photosphere, occurring during the magnetization process, tends to prevent very high intensities from being built up at the photosphere. The absence of sunspots in high latitudes can be explained by the following *ad hoc* postulates:

(*a*) $$30^\circ < \lambda < 90^\circ.$$

The convective motions are strong enough to prevent the formation of regions of high magnetic intensity.

(*b*) $$\lambda < 30^\circ.$$

The convective motions are not strong enough to prevent regions with magnetic intensities $\sim 10^4$ gauss from being occasionally built-up.

The perturbing effect of the major planets, and of Jupiter in particular, may lead to an important contribution to $\boldsymbol{u}$. For example, the perturbation in the velocity of material at a distance from the sun comparable with the radius r_J of Jupiter's orbit is $\sim(m/M) \,.\, (GM/r_J)^{\frac{1}{2}}$, where m is the mass of Jupiter. Since r_J is of the same order as the accretion capture radius, the main perturbation due to Jupiter takes place at distances where the distribution of accreted material is not spherically symmetric (see section 10). It can then be shown that the accreted material acquires a non-zero average angular momentum per unit mass (the average being taken with respect to both spacial position and time). This angular momentum, which is of the order of, but less than, $(mr_J/M) \,.\, (GM/r_J)^{\frac{1}{2}}$, is preserved as the material falls towards the sun. Accordingly, the material enters the solar atmosphere with a systematic average horizontal velocity of the same order as, but less than, $(mr_J/MR) \,.\, (GM/r_J)^{\frac{1}{2}}$. This velocity is directed in circles about an axis perpendicular to the plane of Jupiter's orbit, and is $\sim(GM/r_J)^{\frac{1}{2}} \sim 13$ km./sec., since $mr_J/MR \sim 1$. The main changes arising from this alteration in the interpretation of $\boldsymbol{u}$ are:

(*a*) The magnetization is somewhat stronger than in (iii),

because $(GM/r_J)^{\frac{1}{2}}$ is greater than the value, 2 cos λ km./sec., used above for $|\boldsymbol{u}|$.

(*b*) The angle between the magnetic axis and the axis of rotation is $\sim 7°$. It is possible that both the absence of sunspots in the equatorial belt between 8° N. and 8° S. and the equatorial acceleration of the sun arises from the inclination between these axes.

The process described above is of general applicability, and in particular gives high magnetic intensities in massive rapidly accreting stars.

(v) *The terrestrial magnetic field*

In section 27 it was suggested that magnetic storms can be explained in terms of magnetic energy transported from the sun by a stream of particles. According to 26 (iv) the intensities of solar flares are such that the ejection of particles is limited to regions where the magnetic intensity is $< \sim 100$ gauss. At times of very large accretion, however, it is probable that no such restriction occurs, for it is to be expected that solar activity would then be on a grander scale than is at present observed. Accordingly, we may suppose that corpuscular streams could be emitted, at such times, from regions where the magnetic intensity attained its highest value. Thus, for a solar magnetic intensity of 17,000 gauss, and taking the field carried by a corpuscular stream as being approximately proportional to $ar^{-3/2}$ (as in section 27), we obtain a storm field at the earth $\sim 0{\cdot}37$ gauss. This value is so close to the present intensity of the Earth's main field as to suggest that the Earth may have been magnetized $\sim 3 \times 10^5$ years ago, during a period in which the rate of accretion by the sun attained a particularly high value.

On this suggestion the magnetic field of the Earth is a transient phenomenon and must evidently be decaying at the present time. The characteristic decay period is $\sim \sigma a^2/c^2$, where a is the radius, and σ is the conductivity of material, in the metallic core of the Earth (the decay time for the outer rock shell of the Earth is comparatively small). It is probable that the effect of pressure in the core is to increase the metallic conductivity to $\sim 10^{18}$. Accordingly, putting $a = 3{\cdot}4 \times 10^8$ cm., the decay time is $\sim 3 \times 10^6$ years.

SUPPLEMENTARY NOTES

27. Magnetic Storms and Auroræ.

It has been shown by J. W. Dungey that the latitude distribution of auroræ can be satisfactorily explained on this theory. Dungey's results depend on the strength H_s of the intrinsic magnetic field carried by the corpuscular beam, and the angle γ between the direction of this field and the Earth's magnetic axis. When $0 < \gamma < \pi/2$ the largest magnetic co-latitudes (both N and S) at which auroræ occur are given for various H_s by the representative values shown in Table 36.

Table 36

$0 < \gamma < \pi/2$

H_s (gauss)	10^{-5}	10^{-4}	10^{-3}	10^{-2}
Magnetic co-latitude	12°	18°	28°	43°

The largest magnetic co-latitude attained (for fixed H_s) decreases nearly linearly to zero as γ increases from $\pi/2$ to π.

The auroræ are distributed asymmetrically in longitude, except in the special cases $\gamma = 0$ and π. The values of γ, H_s may vary with time during a magnetic storm. Such variations lead to corresponding variations of latitude and longitude.

Although a general correlation is to be expected between H_s and the observed geomagnetic storm intensity, the correlation is not a strict one, since the observed field is dominated by the local effect of currents induced in the corpuscular beam through the action of the Earth's magnetic field. Such induced effects are negligible, however, at the neutral points where the acceleration of the auroral particles takes place. It is seen therefore that, since auroral activity depends on the intrinsic field carried by the beam whereas the observed geomagnetic storm field depends mainly on secondary induction effects, there can be no strict correlation between auroral activity and the observed strength of a geomagnetic storm. This is in accordance with observation which shows only a general correlation.

28. The Ionosphere.

Recent work by D. R. Bates suggests that the observed intensity of the ' dawn flash ', arising from scattering of solar radiation by N_2^+ in the Earth's atmosphere, may be in conflict with the high energy photon theory of the formation of the E-layer. An alternative theory of the formation of this layer, depending on pre-ionization of O_2 by quanta with energies slightly above 12·2 eV. has been advanced by M. Nicolet.† A decision between the two theories awaits new observational material.

29. Emission of Radio Waves under Thermodynamic Conditions.

An important paper by D. F. Martyn has recently appeared.‡ Martyn applies ionospheric theory to the solar atmosphere and concludes that reflection plays an important rôle at wavelengths greater than ~ 2 metres. The reflection process arises from scattering by free electrons. At first sight it would seem that such an effect must be quite unimportant (compare with the scattering of visible light by the corona), but account has to be taken of coherence which greatly increases the scattering at radio wavelengths.

Martyn finds that, if the radio frequency is appreciably greater than the geomagnetic frequency (the case discussed in section 29 (i)), reflective effects reduce the effective temperature at wavelengths $> \sim 2$ metres. There is a reduction by a factor ~ 2 at 5 metres. Reflection is less important in the case of magnetic resonance (sections 29 (ii) and (30)).

30. The Origin of the Magnetic Field of the Sun.

P. A. Sweet (paper in the press), using a different method, has independently arrived at the conclusion that the accretion process is responsible for the magnetization of the sun.

† M. Nicolet, *Mem. R. Met. Inst., Belgium*, **19**, 1–162, 1945.
‡ D. F. Martyn, *Proc. Roy. Soc.*, **193**, 44, 1948.

Zürich Meeting of the International Astronomical Union.

The following matters were raised at the Zürich meeting of the International Astronomical Union.

M. A. Ellison pointed out that a quantitative method for determining the importance of a solar flare is urgently required. The rough method, based on area, at present used leads to large discrepancies between the results presented by different observatories.

According to D. H. Menzel it is doubtful whether significance can be attached to observations of calcium ions travelling from the sun to the Earth (see page 39).

W. O. Roberts reported a marked weakening during the last two years in the correlation between regions of bright coronal emission and disturbances of the geomagnetic field. Such a weakening is to be expected, on account of the increased geomagnetic effects arising from particles emitted during solar flares.

INDEX

Printed in the United States
By Bookmasters